D7:1977

PLANNING OF SUBSURFACE USE

Birger Jansson
Torbjörn Winqvist

This document refers to research grant 730591-8 from the Swedish Council for Building Research to VBB,Vattenbyggnadsbyrån, Stockholm.

Permission for distribution approred by the Sec[illegible]

The National Land Survey of Sweden 1975-05-2[illegible]

Distributed throughout the world by
Pergamon Press Ltd.

U.K.	Pergamon Press Ltd., Headington Hill Hall, Oxford OX3 0BW, England
U.S.A.	Pergamon Press Inc., Maxwell House, Fairview Park, Elmsford, New York 10523, U.S.A.
CANADA	Pergamon of Canada Ltd., 75 The East Mall, Toronto, Ontario, Canada
AUSTRALIA	Pergamon Press (Aust.) Pty. Ltd., 19a Boundary Street, Rushcutters Bay. N.S.W. 2011, Australia
FRANCE	Pergamon Press SARL, 24 rue des Ecoles, 75240 Paris, Cedex 05, France
FEDERAL REPUBLIC OF GERMANY	Pergamon Press GmbH, 6242 Kronberg/Taunus, Pferdstrasse 1, Federal Republic of Germany

First edition 1977

ISBN 0 08 022689 2

Key words:

rock cavity
tunnels
subsurface
urban planning
responsibility of planning
building control
inventory of establishments

UDC 69.035.4:711
624.19
69.009

D7:1977

ISBN 91-540-2707-1

Statens råd för byggnadsforskning, Stockholm

LiberTryck Stockholm 1977

FOREWORD

Background to the project

The technique of subsurface construction in the form of tunnels and rock cavities has been developed in Sweden to a very high level. The extent of subsurface construction work is considerable compared with that in other countries, both in relation to the size of population and also in absolute figures. The utilisation of cavities below the ground has developed in a very short time to cover a large number of individual activities.

As a result of construction, skill and a wide range of knowledge concerning the characteristics of subsurface installations, as well as of geological and hydrological characteristics, it has been possible to build subsurface installations in a very satisfactory way from the standpoint of function and economy. There has been very little attention paid from the outside to subsurface construction and to subsurface utilisation. As a result of this, skill and experience have not been exploited sufficiently, so that not infrequently damage in the ground and to buildings occurs as a result of the lowering of the ground-water table, landslides and settling. During both the construction and utilisation stages, there is also occasional disturbance caused by noise and vibration, which is undesirable for the environment.

Certain subsurface areas in large towns are attractive for different types of activity and in time become so riddled with installations close to the surface of the ground that crowding and conflicts occur. Each special requirement for space is not infrequently satisfied in a special tunnel often parallel to tunnels for other purposes. Unnecessary cost and the spread of disturbance or damage can result from the fact that different combinable demands are not co-ordinated within the same space.

In certain cases, activities may be forced to use subsurface installations in spite of the fact that location above ground level would be preferable. In other cases, the possibility of locating activities below ground level has not been utilised due to a lack of knowledge about subsurface construction technology or due to insufficient analysis.

These examples of lack of foresight in the utilisation of subsurface areas implies the risk that greater conflicts may occur in the future in connection with increased utilisation of subsurface areas. Lack of confidence concerning special subsurface construction technology may develop and the result may be poor utilisation of the subsurface as an important resource in town planning.

There is an obvious demand for better exchange of knowledge and better conditions concerning the utilisation of subsurface areas. More attention must be paid to this subject by authorities, planning staffs, decision-makers and legislative experts. Methods for and the organisation of subsurface installation planning should be examined.

The aim of the project

The general background described above gives the motivation for development of the planning of subsurface installations to a satisfactory level. For this reason, financial allocations have been applied for and granted by the Swedish Government Council for Building Research for this research project, which is known as "Planning of Subsurface Use".

The work of the research project has been to obtain an overview of all the factors influencing the planning of subsurface installations and to present a comprehensive viewpoint for this field in order to help those responsible for an installation to understand the demands made concerning foresight and co-ordination, and also to draw the attention of the authorities to the part played in the community by subsurface construction. It has not, on the other hand, been the intention to carry out more intensive studies within the fields concerned. The project has been primarily intended to collect and compile the range of knowledge and experience available among those active within subsurface construction, from authorities and from other parties concerned.

The intended direct result of the project was specified in the following points:

- contributions with opinions on current work on modified building legislation within the Housing Department,
- contributions with opinions on the work and instructions for planning and construction within the Swedish Board of Urban Planning
- basic material for research and development work within the field of knowledge concerning planning,
- basic material for a continued international exchange of experience.

As the work has proceeded, it has been found that there is a clear need of information and reference material among authorities and institutions and also among other people active in this field. Since the survey material has been found to satisfy this demand to a certain extent, it has been included in the summary.

The layout of the project

The project has been limited mainly to events and problems in connection with existing and future towns and does not therefore cover problems involved in the planning of mines. However, due attention has been devoted to the extensive experience in the mining field of problems similar to those occurring in subsurface construction.

The project has been divided into a number of main parts covering the entire range of products, which makes possible a dividing up of the survey work between those taking part. The figure shows the main parts of the project.

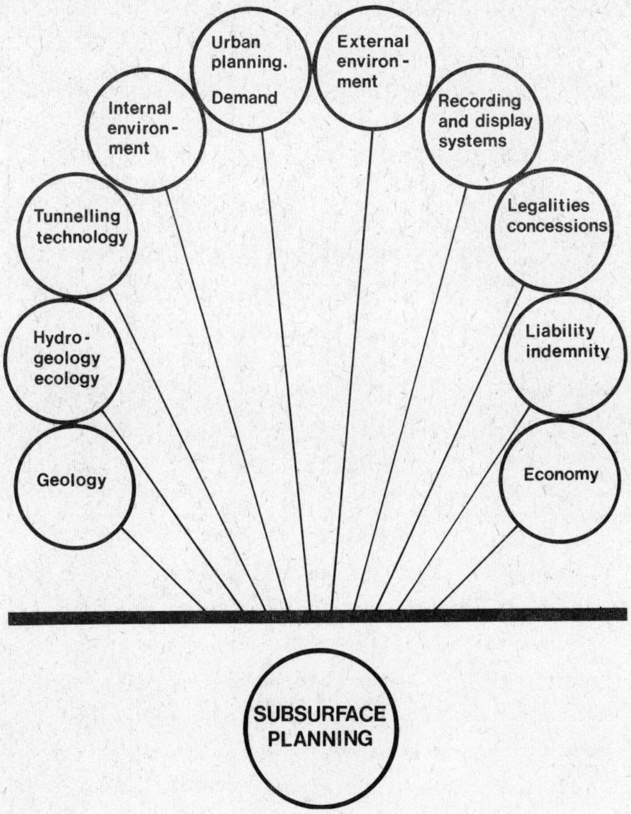

The project has been divided into four stages:

Stage 1 (December 1973 - March 1974), consisting of programme work and precise definition of the project.

Stage 2 (December 1973 - October 1974), consisting of preparatory inventories, selection and preparation of the sample planning areas.

Stage 3 (May 1973 - January 1975), consisting of inventories, analysis and development of planning methods, sample planning.

Stage 4 (February 1975 - February 1976), consisting of summaries and recommendations, preceded by the distribution of preliminary results for opinions.

Those active in the survey work

The survey work has been carried out by a working group made up of representatives from different technical fields:

Birger Jansson, Architect, VBB (Project leader)

Torbjörn Winqvist, Graduate Engineer, VBB (Secretary)

Rolf Brinck, Graduate Engineer, Skadekonsult AB

Ulf Kihlblom, Doctor of Technology, VBB

Nils-Eric Lindskoug, Graduate Engineer, Sven Tyrén AB

Christer Fritzell, Architect, of the Royal Institute of Technology in Stockholm, has carried out a separate inventory of the range of knowledge available concerning the reaction of human beings to the special subsurface environment during construction work and operation.

Very important contributions have also been made by Harald Berg, Licentiate of Engineering, Stellan Cyon, Architect, Per Eckerborn, Graduate Engineer, Hans Edberg, Technologist, Ulf Ranhagen, Doctor of Technology, all of VBB, and also Hasse Tullberg, Bachelor of Law, Skadekonsult AB.

There has been joint consultation in certain geological matters connected with construction with Hagkonsult AB (Lars Lundström, Graduate Engineer).

During the survey work, the Working Group has been supported by a Reference Group. The members of this have been Sten G A Bergman, Doctor of Technology, Torsten Egerö, Architect, Technical Director Gunnar Hult, Departmental Head Peter Norrman and Director Sven Platzer.

Special contact has been established with two of the people for whom the survey is mainly intended, namely the Senior Secretary of the now dissolved Committee of Inquiry into the Building Laws, Bengt Hedman, Head of Legal Section of the Housing Department, and Peter Heimbürger, Head of Planning Section at the National Board of Urban planning.

A number of other people, mainly within VBB Vattenbyggnadsbyrån, have been engaged to a greater or lesser extent.

The execution of the project

The research project has been carried out in various respects in a different way from that originally planned. Since inventory was carried out during the two preparatory stages, it was found that the information about the various survey areas was considerably more useful than had been originally supposed. The sample planning section was therefore abandoned in favour of more intensive surveys and analysis of selected areas. It was later found important to carry out an extensive inventory of information and reference material. These two procedures resulted in a further extension to the scope of the project, Stage 3 being sub-divided into two sections.

It was presupposed that a study on the reaction of human beings to underground environments would be carried out by a group of nine experts supported by the special survey secretary. During the execution of the project, it was found that the members of the expert group could not provide sufficient of their own work. For this reason the survey was carried out by the Secretary, Christer Fritzell, supported by Professor Harriet Ryd, of the Royal Institute of Technology in Stockholm.

The preliminary conclusions and the most important sections of the back-up surveys from the first three stages were summarised in the form of a report called Final Report 3 (1975-04-18). This report was distributed for comment to authorities, state and local government departments, institutions, universities, organisations within the branch, and to contracting, industrial and consulting companies, making up a total of 100. The number of responses received was 60, 50 of which must be considered as being well or very well considered. This response from those involved in subsurface activities must be considered as an extremely good one.

This report consists of a revised version of the preliminary edition which was issued for comment. Due attention has been paid to contributions and viewpoints from the sources referred to. The earlier layout, which was divided up for practical and financial reasons under different authors, has been revised to a layout which is based more on results and which is easier to read.

Acknowledgements

This research project has only been possible through the solid and generous support of a large number of authorities, companies and private people. It has not been difficult to obtain the assistance we needed. We believe that this is associated with the fact that, even if opinions vary concerning our work, the importance of the project and its good intention have never been queried.

Contact has been established with a very large number of people as the survey has proceeded. You have all been more than helpful. We should particularly like to mention the local authorities in Gothenburg, Stockholm and Uddevalla who have made the in-depth surveys possible.

The bodies referred to for comment and opinion received material which was both comprehensive and in several respects very difficult to tackle. The imposing amount of work carried out has considerably increased the value of this report.

We should also very much like to mention all those who, inside and outside the Parent Company, have willingly contributed their expertise during the various stages of the project.

Our grateful thanks to you all.

Stockholm in August, 1977

Birger Jansson Torbjörn Winqvist

TABLE OF CONTENTS

1. INTRODUCTION

Society makes demands on activities within the construction field.

- The resources of society are to be safeguarded in a sensible way;
- collective needs are to be satisfied;
- the life and property of the individual citizen are to be protected.

It is difficult to find any reason why subsurface construction should be regarded in a different way from any other type of construction. Nevertheless, the creation and operation of subsurface installations have failed to attract anything like the amount of attention given to surface construction.

1.1 Limitations and definitions

The aim of the research project is to provide a picture of those aspects of the work which are relevant from a construction point of view, and from that of society, i.e. the user. Demands on the process of creation are formulated and proposals are made as to how these can be satisfied. Only those characteristics which distinguish subsurface construction from other types of construction have been treated.

The most important of these are:

- Exploitation of the characteristics of the surrounding soil and rock with respect to the stability and use of the rock cavity.
- A special construction technique for the creation of rock cavities.
- The influence of the cavity on the surrounding mass of soil and rock and on the ground-water.
- Rock cavities are of a permanent nature and can be altered only with very great difficulty.
- Rock cavities are dark spaces with only limited access to the surface.
- The creation of rock cavities results in surplus excavated material corresponding to the size of the cavity.

It is not the aim of the project to deal with subsurface construction and installations which can be assigned to conventional surface construction i.e. cellars, basements and foundations to a depth of one storey, nor with the construction of conduits of small dimensions at shallow depths.

One attempt to define the concept "subsurface construction" which is applicable to this project was made in preparation for the OECD Advisory Conference on Tunnelling held in Washington D.C. in 1970. Artificial cavities below the surface in rock, soil or water having an area in cross section of 2 m^2 or greater and which are not subsurface parts of surface installations are classed as subsurface installations.

The criterion of size should be the guiding principle even if cavities of smaller dimension possess all or most of the features characteristic of subsurface construction.

The dividing line between subsurface and surface installations should be drawn so that those installations in the subsurface possessing all or most of the features characteristic of subsurface installations are regarded as subsurface installations even if they are connected to surface installations.

The research project does not deal with subsurface construction aimed at the extraction of minerals, which may be referred to mining operations.

2. CHARACTERISTICS OF SUBSURFACE CONSTRUCTION

The following pages provide an introduction to the various factors and aspects which characterise subsurface installations and subsurface construction. The aim is to supply the reader with information regarding these special traits and to facilitate understanding of the subsequent presentation.

2.1 Geology imposes governing conditions

Earth and rock surrounding a rock cavity are suspended by an arch which is formed in the surrounding mass of rock. The arch becomes active through a structural deformation which takes place during the excavation. Normally, the active part of the rock is considerably larger than the cavity itself and distortions are small - on the condition that the rock quality is good. The important properties, from the point of view of construction are: good compressive strength, low frequency of fissures and fine cracks and absence of clayed deposits, crush zones and other weak zones.

No rock mass is ever homogeneous and certain imperfections must be accepted. Rock is composed of a great number of different types of material each with its own particular qualities. As a result of movements in the earth's crust (tectonic activity) the bedrock is criss crossed by crush zones and fissure systems of varying magnitude. Close to the surface the rock is more broken and its suitability for building purposes can vary considerably from place to place. At greater depths the rock becomes more compact, but at the same time rock and water pressures increase. During the planning stage the nature of the local rock mass is taken into consideration by, for example, locating the cavity in relation to the rock's fissure system, when working out detailed cross-sections, etc.

The possibilities of evaluating the geological conditions for a subsurface system are limited as soon as one moves away from the

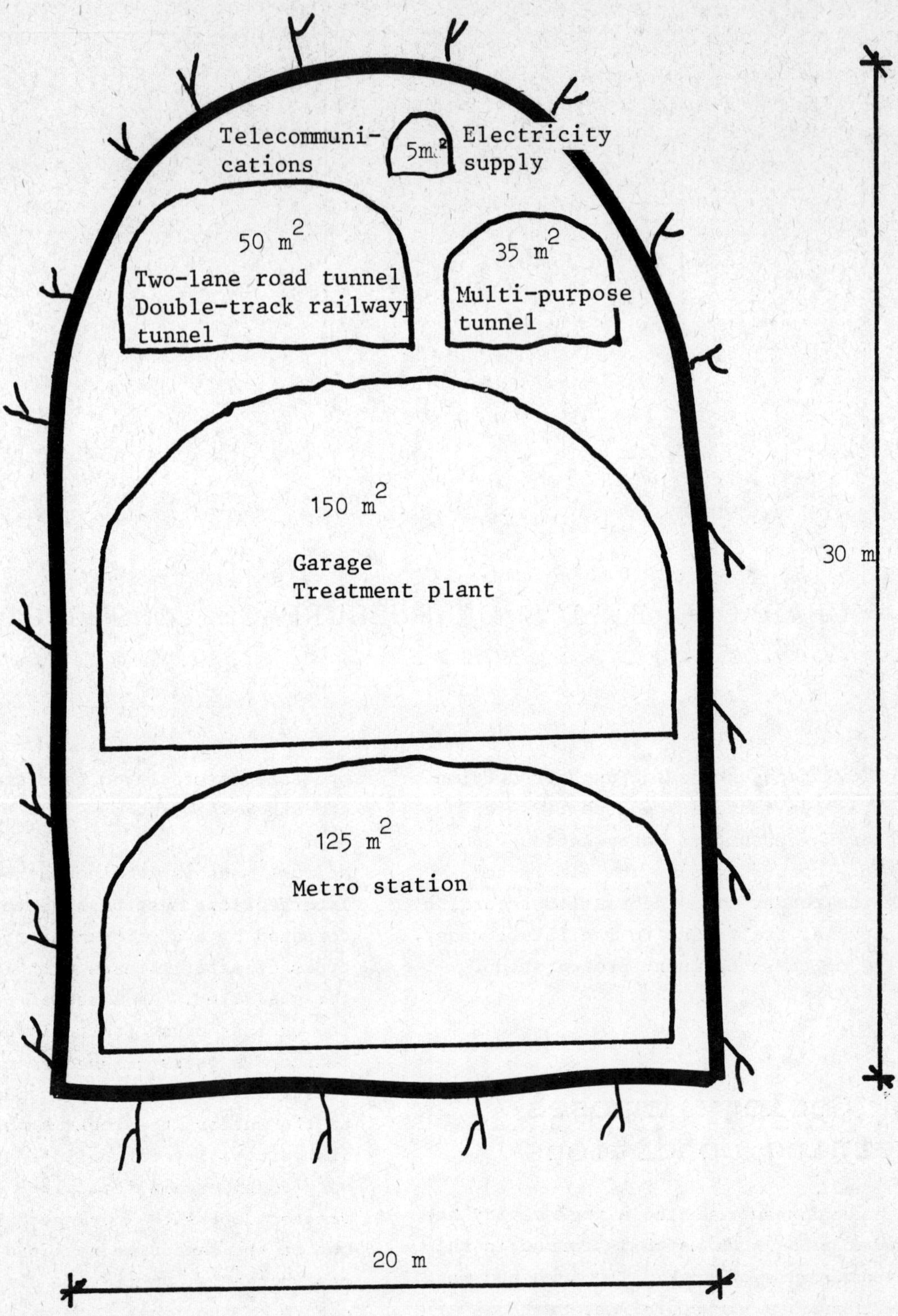

Figure 1

Illustrated above are rock installations for various purposes, shown with typical sectional areas. The surrounding, rough framework refers to rock cavity for oil storage.

vicinity of the surface, due to the impermeability of the ground. Overall evaluations can normally be made with a considerable amount of accuracy and at a fairly low cost by studying the geology with the aid of aerial photographs and by carrying out aerial magnetical surveys etc. More accurate surveys which demand work on the ground, such as various types of test drilling, are both extensive in nature and expensive to carry out. For this reason there are seldom either technical or economic grounds for trying to gather absolutely accurate information for a project. That is why subsurface construction is always characterised by a certain degree of uncertainty.

In comparison with other countries, Sweden is endowed with good quality rock which is located near the surface due to the fact that poorer quality, weathered rock has literally been scraped away during, if looked at in geological time, recent periods of glacial activity. The bedrock consists to a large extent of primary rock possessing qualities which are particularly advantageous for subsurface construction. Rock types with generally poorer qualities from the point of view of construction are the slates found in the mountainous regions and the sedimentary rocks found for example in Scania, Öland and Gotland. Considerable local variations in structure and strength can occur.

2.2 Excavation techniques

Drilling/blasting is by far the dominant method of construction in Scandinavia and is likely to remain so for a good many years. Under favourable conditions large rock structures with dimensions of more than 20 m wide and 30 m high are currently and regularly being constructed. Research work is being carried out with the object of eventually doubling and maybe even trebling the present spans. The cavity can be made to any required length, given that the rock quality is satisfactory. There are examples of power station tunnels many miles in length with considerable areas - anything up to three hundred square metres or more - despite the fact that the rock has not always been of the highest quality.

The rock cover should generally speaking always be at least the same size as the span. Adjacent cavities must be separated from each other by walls or columns so that the loadbearing arch can be formed. If the structures are located at different heights, consideration must be given to the arches' - the compressed zones' - distribution above and below the respective cavity levels.

Normally the rock has to be reinforced to a certain extent to provide protection against falling stones and to make good local anomalies in the rock. Should there be any danger of harmful ground-water leakage occuring, precautions have to be taken to make the defective section watertight. As the rock quality decreases so the need for reinforcement and sealing measures increases at the same time as building scope is reduced. From the point of view of cost and time, extensive reinforcement work can be more demanding under unfavourable conditions than the actual excavation operations.

Under normally good rock conditions, narrow tunnels with cross-sectional areas of up to 12 m^2 can be driven at a speed of 35-40 m per week and front on a 2-shift system. One tunnel 100 m^2 in size can be driven 15-20 m per week by the same number of workmen. The average rate of progress for large cavities can thus be given as 5-10000 m^3 or more per week.

The costs of construction and reinforcement work are given in section 9.5.

Other construction methods are employed when the demand for caution is great or when construction work is being carried out either in earth or in rock with low strength qualities. In the latter cases static equilibrium must normally be maintained with the aid of load-bearing structures, i.e. some form of total encasement. Construction techniques become more complicated and costs and amount of construction time required increase.

2.3 Utilisation of the properties of a cavity

The term subsurface construction implies that a cavity is created in a finished building material. The existing rock or earth masses provide the subsurface cavity with different characteristic properties. These properties are in many respects valuable and justify the installation of various types of plant under the ground. A brief summary is given below.

Protection against climatic influences

Subsurface space is not affected by precipitation, wind conditions, air humidity or temperature. Exceptionally good heat insulation and heat capacity characteristics mean that high or low temperatures can be maintained at a low level of energy input. Furthermore, conditions are ideal for effective climatic regulation.

Protection against damage (accidental or deliberate)

The element of risk involved in surface construction, for instance when pipes are dug up, when disasters occur in the immediate vicinity etc., is reduced or completely eliminated. Protection against sabotage and the effects of war is improved.

Increased accessibility

For various types of operation the possibilities for inspection and maintenance are improved through subsurface installation as compared with traditional surface-level techniques. Examples in this connection are various types of pipe which are laid in pipe tunnels instead of being buried under the ground.

Protection from observation from outside and from disturbance

Cavities are protected from observation from outside and from noise, vibration, pollution etc., which are caused in the surroundings. Conversely, disruptive or dangerous operations can be effectively screened off from the outside world through subsurface installation.

Consideration of the environment, in the form of good visual surroundings, freedom from noise, clean and fresh air, and conservation of natural and cultural environments, can be achieved by locating to the subsurface unsavoury, disturbing or in some other way unattractive activities.

2.4 Exterior factors affect subsurface construction

In addition to the properties which the surrounding medium furnishes a subsurface cavity with, the attraction of subsurface installation depends on a number of external factors. Their influence varies as much with the type of operation proposed as with the prevailing local conditions.

Topography

Undulating terrain with hills, valleys and steep slopes can both create a need for and simplify subsurface location. Subsurface location can be advantageous where ground conditions are difficult (boulder-strewn terrain, poor load-bearing qualities etc.) and in the vicinity of water courses.

Figure 2

To improve the traffic situation in a part of Gothenburg a new highway is being considered. As a surface position would imply a considerable encroachment on the hilly country and, not least, on the area of the botanical gardens, alternatives, either partly or wholly subsurface, are being considered (black dashed marking).

Gothenburg Town Planning Department

Utilisation of natural resources

Natural resources on the surface, including unspoilt land, can be conserved if future installations are located to the subsurface. Conservation of energy can be improved by locating installations under the ground (for example, high or low temperature storage, storage of energy in the form of rock accumulators, hydro-electric pumped storage, compressed air storage). Unspoilt land can be preserved for other purposes, which is of particular importance in urban areas. When earth and rock masses from the construction of cavities are preserved, the demand for exploitation of natural resources of gravel is reduced. (Subsurface projects should therefore come under a general plan for surplus material utilisation).

Urban structure

High urban density, large urban areas or agglomerations are usually characterised by a low degree of flexibility in land utilisation and land resources. Additional demands for space for some form of activity can be net only at the price of serious construction problems or, alternatively, result in selection of a location far from the town centre. The alternative subsurface location can frequently make available sufficient space located in the required position. Installations, such as metros, road tunnels, pipe and cable tunnels for electricity, telephone, district heating, etc., extending over several or many properties can be constructed without coming into conflict with surface-level buildings. Installations which, if located at surface level, would cause severe disruption during the construction phase, can be constructed without the occurrence of such problems if they are located under the ground.

Society's economic and social standards

It can be assumed that the tendency towards locating activities under the ground is closely connected to society's economic and social standards. Environmental consciousness and demands for high environmental standards lead to the allocation of economic resources to installations which effectively satisfy these demands. Technical solutions which give long-term economy and high functional efficiency are also favoured. The establishment of installations for economic and military defence readiness is also guided by criteria connected with standard and resources.

Technical level

The cost relationship between the different alternative ways of realising of a project in construction and operation is a central factor in decision-making. Through rapid technical development subsurface construction technology has steadily become more and more competitive vis-à-vis corresponding construction elsewhere.

Increases in costs in subsurface construction are relatively slower than in other types of construction. A continuation of this situation should increase the attractiveness of subsurface location as compared to other alternatives.

Disposal of rock or earth masses

The rock or earth masses corresponding to the volume of the excavated cavity must be disposed of. Ideally they should be made use of, for example as crushed material or for some other similar purpose. On the other hand, their disposal can create problems as far as cost and the environment are concerned. This factor has often been decisive as far as the economic outcome of a project is concerned. Such surplus material can be made use of as building material in crushed or uncrushed form, or as fill, for example in the construction of embankments, harbours etc.

2.5 The subsurface environment

The location of certain activities under the ground means that people have to stay under the ground for both shorter or longer periods of time (e.g. journeys by metro, industrial operations). The subsurface environment possesses a number of distinctive features:

- lack of windows
- limited possibility of evacuation
- artificial climate
- special room characteristics, especially in the case of unlined rooms
- in most cases, vertical exits
- rock and/or earth masses overhead

The way in which people experience being under the ground varies from individual to individual and depends as much on the surroundings as on the conditions under which the visit takes place. How much choice there is in being under the ground, the length of the visit, the type of activity, the opportunities for contact with other people, and the degree of actual and experienced safety are all examples of such conditions.

2.6 Environmental impact

The drilling/blasting method, which is the method most often employed in hard rock, gives rise to vibrations, and neighbouring buildings can be damaged. It is, however, possible to reduce the effects of this disturbance.

Techniques for reducing disturbance is, highly developed and detailed regulations against disturbing vibrations are as a rule applied.

Subsurface construction entails the necessity of transporting away surplus material which has been excavated. These heavy and often frequent transport operations can result in disturbance to the surrounding area. The depositing of large quantities of surplus material can lead to considerable encroachment on the ecology and surrounding scenery.

Leakage of ground-water during the construction and operational stages can give rise to changes in ground-water pressure and drawdown of the ground-water table. A lowering of the ground-water table has a varying degree of influence on the aquifer as a whole, even though the most pronounced effects are frequently only local. With good hydraulic communication, and this should be particularly emphasised, it is possible for the effects to extend over great distances (many hundreds of metres).

The consequences of changes in ground-water pressure can be that:

- Foundations resting on wooden piles or mattresses (usually under older buildings) tend to rot on contact with oxygen in the air, which either damages or destroys the building in question.
- Soil settlement occurs in compressible layers of earth. This can lead to the buildings and installations affected being seriously damaged. Soil settlement takes place over a long period of time (10-100 years) with decreasing rapidity and can, for example in loosely packed clay, amount to a tenth of the total thickness of the clay layer. The course of the settlement can, under favourable conditions, be stopped but never reversed.
- During the operational stage leakage of stored or transported media as heat or cold substance can affect the surroundings.
- Conditions underlying the ecological system (in terms of animal and plant life) are altered.

2.7 A rock cavity cannot be removed

A cavity excavated from the rock can scarcely be reinstated. It could conceivably be filled with concrete but this is regarded as being difficult to carry out. It should therefore be assumed that cavities constructed in rock are permanent features. In other words, the planning horizon for subsurface structures is infinite.

In this context, consideration must not be limited to individual cavities only but must encompass the stability of the rock mass in its entirety. The proximity of a cavity restricts subsurface utilisation. Superficial rock resources in attractive positions should therefore be made the object of careful planning, in which the needs of different functions in the long and short terms shall be taken into account.

2.8 Subsurface construction is three-dimensional

Construction at ground level is regulated by detailed physical plans. The planners have taken into account both topography and ground conditions as well as the existing buildings, the possibility of constructing access roads, water supply and sewerage facilities etc. Each individual building starts off at ground level - a surface - and can be extended upwards (several storeys) and in certain cases, downwards (several basement floors).

Subsurface conditions are radically different. The subsurface must be regarded as being more or less unexploited. Building scope is mainly limited by the geological structure of the ground and the demand for stability. The significance of this can be illustrated as follows:

- Subsurface installations can be built to great depths
- They can be designed with a large amount of freedom vertically and laterally.
- Different subsurface installations can be constructed below (or above) each other.
- Adjacent subsurface installations are to a large extent independent of each other.

Up to now subsurface installations have been located relatively near the ground surface. Any further extension is downwards in the form of a second and third "underground storey" etc. Looked at in a wider perspective, the entire subsurface area can be seen in the form of a 3-dimensional model with installations positioned in the available space and the ground surface serving as the only limiting factor.

2.9 Right of disposition

Formally, the land owned by a property owner extends from the ground surface to the centre of the earth. In practice, however, the right of ownership is limited in the case of greater depths. Tunnel owners can be given the right to build through somebody else's property and pay very little or no compensation. The laws governing property ownership are often considerably less complicated when it comes to subsurface construction than they are for ground level construction. In other words, it can prove to be cheaper, simpler and quicker to obtain the right of disposition below the ground surface than is the case for construction at ground level.

2.10 Interaction between surface and subsurface

Subsurface installations have both a physical and a functional connection with the ground surface and the conditions prevailing there. In densely built-up areas these are pronounced.

The geological and hydrological conditions as well as the presence of previously constructed installations determine the feasibility of subsurface construction. Subsurface installations must be constructed so that the surroundings are not affected, for exemple by buildings being damaged or by the stability of the ground being jeopardized. If various types of activity can be located under rather than above the ground it obviously means that much greater freedom can be given to the way in which the ground surface is used.

The construction of subsurface installations is thus regulated by a mutual, close and often extremely strong connection between the conditions above and below the ground surface. Herein lies the motive for dealing systematically with these clearly differentiated areas of construction in one and the same context.

As the amount of subsurface construction increases, connections with the ground surface will be of particular interest. Since the geological conditions for building are normally worse near the surface at the same time as utilisation of the ground surface often results in restrictions, it can be expected that it will be here that conflict and crowding will initially occur.

Areas of use currently located in the subsurface

Areas of use	User			Geografical distribution	
	State	Local Authority	Individual	Urban	Non-urban
TRANSPORTATION					
Supply tunnels with one or several functions	X	X		X	
Fresh, waste-, storm- and heated water (district heating)					
gas					
electricity and telecommunication cables					
pneumatic waste disposal tubes					
Traffic tunnels					
Road traffic	X	X	X	X	X
rail traffic (railway and underground railway)	X	X	X	X	X
pedestrians	X	X	X	X	X
(navigation)					
Industrial supply tunnels					
Oil pipe lines	X	-	X	X	X
bulk transport/ part load transport	X	-	X	X	X
raw water (process water, for water treatment plants)	X	X	X	X	X
cooling water					
STORAGE					
Garages	X	X	X	X	-
silos	X	X	-	X	X
cold store and refrigerated chambers	X	-	X	X	-
archives	X	-	-	X	-
petroleum products	X	X	X	X	X
- crude oil					
- fuels					
- liquid gas					
compressed air	X	-	X	X	X
PRODUCTION AND PROCESS INSTALLATIONS					
Factories	X	-	X	X	X
steam power stations	X	X	X	X	-
nuclear power stations	X	-	X	(X)	X
sewage treatment plants	-	X	-	X	-
freshwater treatment plants	-	X	X	X	X
quarries	X	X	X	(X)	-
hydro-electric power stations	X	-	X	-	X
district heating plants	-	X	-	X	-
TERMINALS					
Transportation terminals, supply	-	X	X	X	-
terminals of goods	X	-	-	X	X
telecommunication centres	X	X	X	X	X
transformer, substations					
DEFENCE INSTALLATIONS					
Military installations	X	-	-	X	X
bomb shelters	X	X	X	X	-
installations of strategic importance	X	X	X	X	X
OTHER POSSIBLE USES					
Assembly and teaching premises	-	-	-	X	-
shopping centres	-	-	-	X	-
sporting facilities	-	-	-	X	-
shooting ranges	-	-	-	(X)	-
burial facilities	-	-	-	(X)	-
waste disposal facilities	-	-	-	-	X

3. OVERVIEW OF SUBSURFACE CONSTRUCTION IN SWEDEN

3.1 Past and future

Sweden has a long tradition of mining, proficiency in the extraction of ore and minerals and familiarity with the subsurface cavities thus created. The need to increase successively the volume of ore mined, as metal and ore, exports have become one of the country's leading basic industries, has provided the stimulus for ever-continuing improvement in mining technology.

This technology has subsequently been brought into use for other purposes when the use of cavities has been able to satisfy needs.

Construction of railways through the hilly landscape required tunnels. Extension of the hydro-electric power potential required water tunnels and that the power house be located underground in rock. From being a primitive art the construction of tunnels and cavities the rock has now become industrialised. This stage can be said to have begun during the Second World War when work on defence and shelter installations reached a high level. Methodology, working cycles and the technology were thoroughly analysed and the experience gained was rapidly transferred to new projects.

The post-war period saw a considerable increase in subsurface construction, which had come into existence to serve functional needs such as tunnels for pipes and cables etc., water treatment plants, metros and the storage of oil.

Informed judgement indicates that the location of society's production and service installations will increase in extent. The reasons for this are the intensive use of land in urban areas, increased environmental demands, but also the technical and economic advantages of subsurface location.

3.2 The extent of subsurface construction

Extent including mining

In 1974, Sweden blasted a total cavity volume in excess of 19 M m³, which from an international point of view is a very high figure. The mining industry's proportion of this was 15 M m³ (80 %).

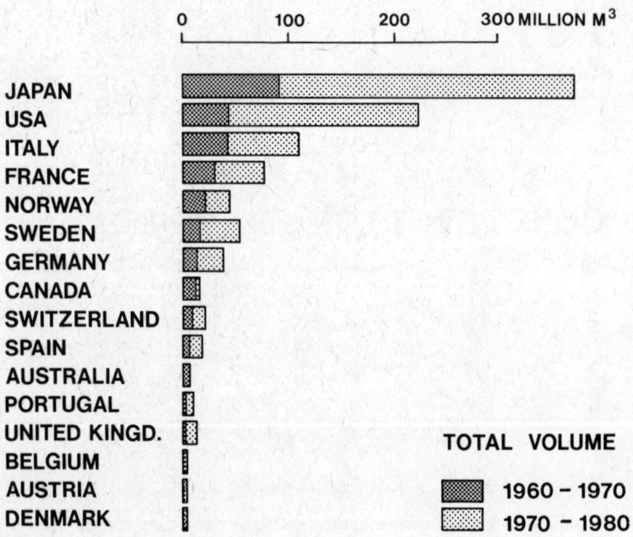

Figure 3
Volume of tunnelling in some OECD countries

If we discount pure mining and only consider tunnels, galleries and shafts of a more permanent character then these still make up a large proportion of total subsurface construction (20 %).

Extent excluding mining

During 1974, it is estimated that approximately 4,100,000 m³ of cavity were excavated at a cost of Sw.Crs 290 M. The total investment during the year including the costs of planning, administration, reinforcement and lining work, etc. are expected to exceed Sw.Crs 1,000 M.

The total production value for subsurface installations totalled Sw.Crs 8,800 M in the same year. In other words, subsurface construction makes up more than 10 % of total construction, excluding buildings.

Internationally we construct, along with Norway, the greatest number of cavities per inhabitant in the world. Construction is almost totally in rock. Only 3 % of subsurface construction is carried out in earth or water. In terms of costs, however, this proportion is larger, approximately 15 %.

Large hydro-electric power stations as a rule include subsurface power houses as well as head- and tailrace tunnels. Thermal power stations located on and under the surface usually include cooling water supply tunnels. In 1974, hydro-electric power stations comprised 50 % of the total volume excavated. Storage installations include primarily storage for oil. These are normally large or very large and relatively cheap. The total volume of storage is increasing at present by 1.5 - 2 M m³ per year.

In 1974, traffic tunnels comprised 5 % of the total yearly volume, which corresponds to 5 km. A large proportion of this length has during recent years gone to the Stockholm metro. In addition, a number of road tunnels and, on odd occasions, railway tunnels have been constructed.

Service tunnels, i.e. tunnels for water, storm water, sewage, district heating, electricity and telecommunications, are constructed to a length of 25-30 km each year. There is pronounced expansion within this sector.

3.3 Current development - forecast

The basic material for statistics and prognoses is limited. In preparation for the OECD Conference in 1970 an inquiry gave a rough overview of subsurface construction in

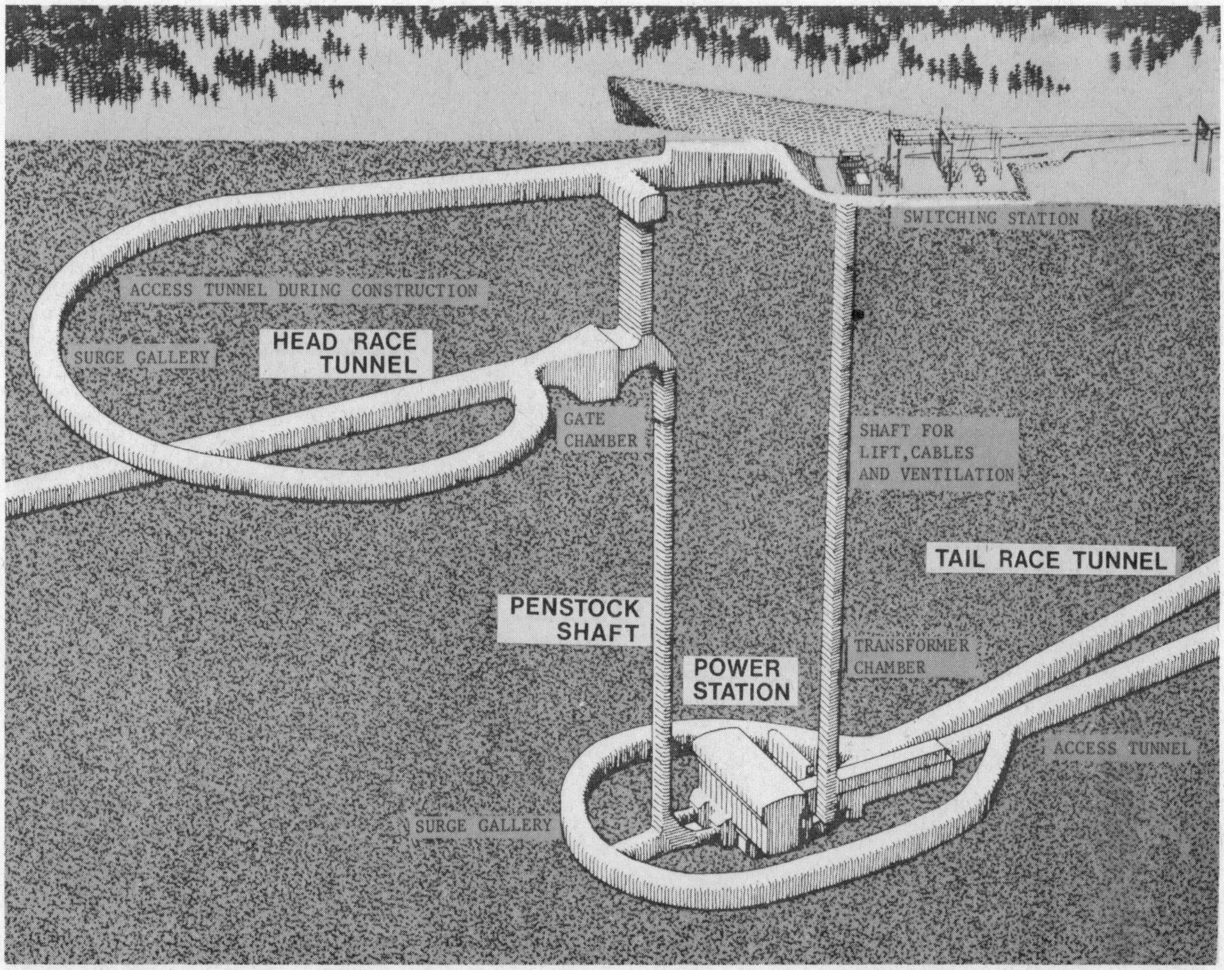

Figure 4

Exemple of a hydro-electric power station (Järnvägsforsen, 105 MW). This type of installation has in recent years corresponded to 50% of all subsurface construction in Sweden with an excavated cavity volume per year of some 2 million m^3.

Vattenbyggnadsbyrån

the 1960's. Amongst other things the aspects of distribution are poorly presented.

The total excavated volume during the 1960's was estimated to approximately 15.5 M m^3. This figure is expected to rise to 38.6 M m^3 during the 1970's, i.e. an increase of 150 %.

In 1974, the picture was complemented by an inquiry carried out by Atlas Copco (Kallin). See table 3:1.

The development which can be observed today (1976) in the form of planned or approved installations indicates that total subsurface construction has reached a peak. Decision to extend hydro-electric power stations and oil storage installations after 1978 has been limited. Regional installations such as the sewerage tunnel systems in Stockholm and Gothenburg, as well as the Stockholm metro, will soon be completed. Of the four sectors, power stations, traffic, service and storage it is only within the service sector that there is evidence to suggest a continued increase in volume. At the same time this sector is the most urban in character and is thus of greatest current interest for this research project.

Figure 5

Oil is lighter than water and does not mix with it. Simply by maintaining the ground-water table well over the level of the stored oil no leakage is possible. Consequently there is no need for sealing measures such as lining etc.

Vattenbyggnadsbyrån

	Result 1974		Prognosis 1974-78	
	Volume Mm^3	Costs mill. Sw Cr	Volume Mm^3	Mean volume per year Mm^3
Power stations	2,04	145	7,6	1,52
Traffic (roads, railways, garages)	0,19	18	0,7	0,14
Service (water supply and sewerage, power, telecommunications, district heat)	0,20	40	1,6	0,32
	0,20	40	1,6	0,32
Storage (oil)	1,67	85	9,3	1,86
Total	4,1	290	19	3,8
Mining industry				
ore and waste rock	15,3			
tunnels, galleries and shafts	1,1			

Figure 6

Excavated volume in Sweden 1974 and prognosis 1974-78

Kallin, Atlas Copco

The limited overview of the subsurface construction field which is available today means that evaluations of the future are difficult. Planning, in the sense this is directed at surface construction, does not exist for subsurface construction. Short preparation time may in many cases characterise subsurface construction projects.

It can therefore be of interest to discuss the development from more general points of departure.

Energy Supply

The principles governing our future energy have not been determined. Studies have been carried out which show potential for further development of hydro-electric installation with subsurface location. It is probable that a continued development of thermal power stations will include subsurface construction. The power stations may to a greater extent be located to the subsurface whether they be oil or nuclear powered. A demand for increased use of district heating in urban areas is expected. Heat production in plants with or without power production requires localisation near urban centres and can result in an increased use of subsurface installations for the plants and for the distribution network.

Oil storage in rock chambers will continue to be constructed on a large scale for the rest of the decade.

A new wave of natural gas storage construction may occur.

Large scale storage of energy in rock chambers in accordance with different storage principles is the object of intensive studies. Examples of such are pumped storage of compressed air, hot water accumulators etc.

Against a background of rising energy costs it can be expected that the excellent insulation properties of rock will attract attention as regards the storage of hot or cold substances, and as regards the reduction in general of energy consumption.

Supply systems in urban areas

Urban service installations such as supply tunnels comprise a small part of the total amount of subsurface construction. Even if such installations have been known since ancient times, greater exploitation has only occurred in the last decade. Continued development can, as has been mentioned, be foreseen. Tunnels of this type can become a natural feature in the exploitation of areas of a certain minimum density. The modification of the present infrastructure in at least the larger existing town centres can result in subsurface construction.

Fixed installations such as sewerage, sub- and transformer stations etc. are already located subsurface today.

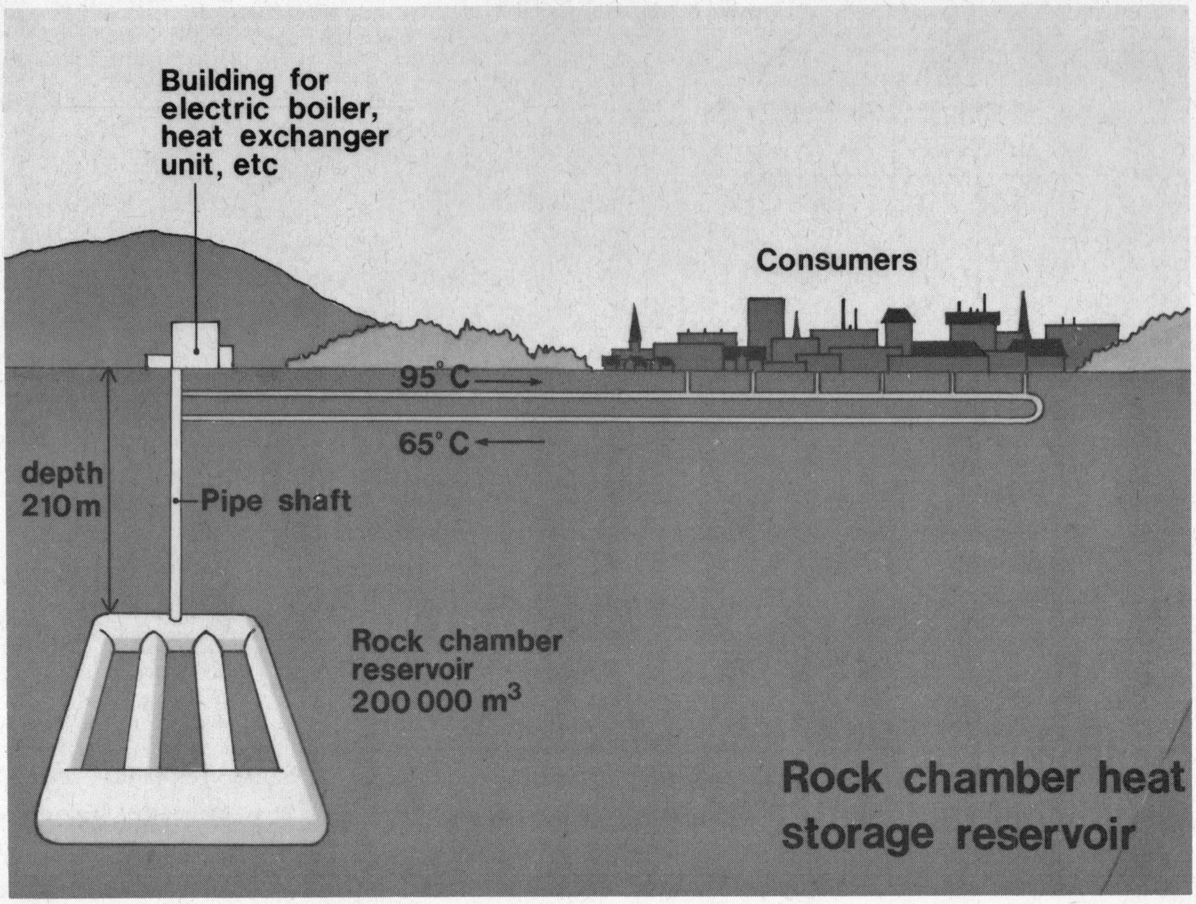

Figure 7
Surplus heating from, for exemple, power production plants can be stored in accumulators deep down in the rock bed for supplying to larger housing areas.
Vattenbyggnadsbyrån

Increased environmental demands, the desire for greater freedom of action on the ground surface, advantageous economic and civil defence aspects justify or alternatively can force an increase in subsurface construction.

Subsurface construction - a resource

The possible different uses of the bedrock as a resource should still be regarded as relatively untried. The physical characteristics of the rock, the characteristic of unused space, i.e. available space, has already resulted in a number of urban subsurface installations such as garages, archives, stone crushers, cold storage.

Installations at deeper levels are completely untried.

It would seem reasonable to forecast an increased interest in finding both areas of use where the characteristics of the bedrock can be exploited, and a more improved technology for the exploitation of this.

4. GENERAL REMARKS ON PLANNING IN SWEDEN

Planning aims at adapting development to the goals set up within the framework of available resources.

Expressed more explicitly this means:

- trying to anticipate furture needs
- designing systems to meet these needs with consideration being given to different kinds of restriction
- securing resources to put these systems into operation

4.1 Levels of planning

Distinction should be made in the following between three levels of planning:

a) planning of individual installations (detail planning)

b) planning of particular functions/sectors (planning of particular functions)

c) planning of the community as a whole (overall planning)

4.1.1 Detail planning

Planning of an individual building or installation aims at adapting the functional demands made on the installation to the physical conditions and to current needs. (For buildings and installations above ground, society requires that the position, use and configuration be closely examined. There is no corresponding general requirement for subsurface installations).

Detail planning is also called design and can be divided into the following steps: preliminary planning, preliminary design, basis for decision, detail design and procurement.

Experts within the field of building and installation construction, together with their client, are responsible for ensuring that the new installation is constructed to meet the goals set up regarding construction technology, economy and function. These experts have only limited responsibility for planning of particular funcions or overall planning.

4.1.2 Planning of particular functions

Different enterprises - which are here understood to include state and local government departments, authorities, private companies and organisations - are responsible for certain amenities or functions in society. Examples of planning of particular functions are the planning of roads, of water supply and sewerage, of electricity and telecommunications, of housing etc. Planning of particular functions aims at meeting the demands made on a particular function in the long and short terms and within the framework set up by society.

Experts within the field of the particular function, together with their client, are responsible for ensuring that the structure of the system and the individual installation satisfy the demands made. It is a requirement that overall planning takes planning of particular functions into consideration.

4.1.3 Overall planning

Overall planning aims at adapting development to the conditions in which society exists and to the goals set up by society for the individual and common good.

Society must weigh up to the demands of different activities at local, regional and national level so that they are kept within the available framework and so that social goals are fulfilled. These aspects which are taken into consideration in overall planning are:

a) economics:

   economic growth, employment, investment, production, maintenance and levels of consumption

b) social structure:

   conditions made by the individual concerning society, health and social services, cultural development

c) physical planning:

   utilisation of land, co-ordination of functions linked to the utilisation of land

d) energy:

   utilisation of resources, conservation, living standards

e) natural resources:

   demands on conservation, reconstruction, ecology

Experts within the various sectors of community planning, together with their superiors, i.e. politicians elected by the people, are responsible for ensuring that indivudual installations and the structures of systems for particular functions can be arranged within society's framework of resources, and that they benefit society and the individual citizen.

In summary, it may be noted that in planning of particular functions statement is made of demands on resources which simultaneously serves as the basis for overall planning. In overall planning, the utilisation of resources is co-ordinated and given limits, providing the conditions for further planning of particular functions and installation planning.

4.2 Time aspects of planning

Planning is carried out over periods of different length.

It is said that the thing most difficult to prophesy is the future. At the same time, demands are made for increased foresight and for the formulation of goals for the future development of society. Attempts to solve this planning dilemma can be made by introducing suitable time limits in different types of planning and by making outline planning an ongoing process.

Time limits can be combined with suitable levels of achievement in terms of the amount of detail and information content.

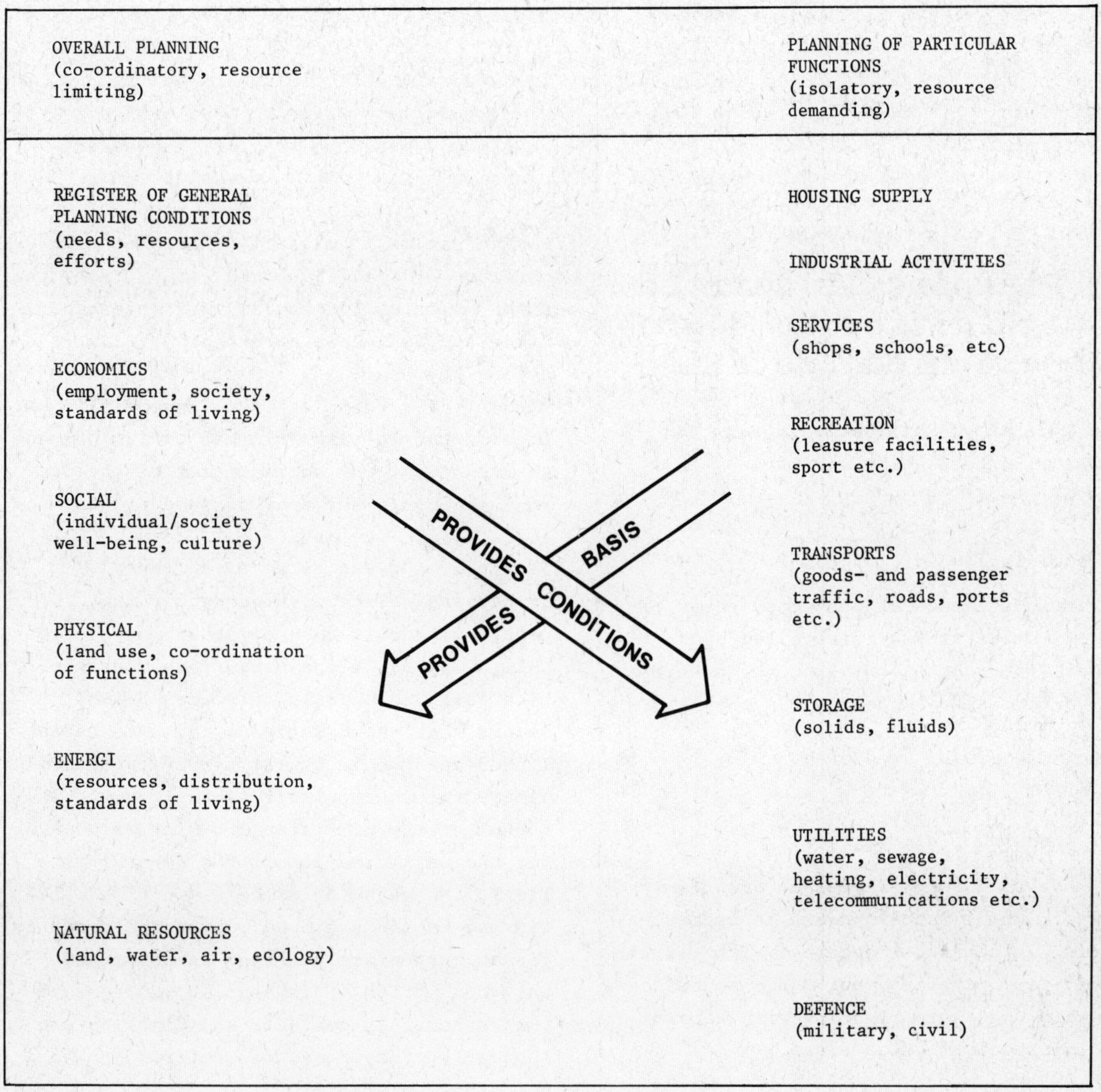

Figure 8

The relation between overall planning and planning for particular functions

Short-term planning - basis for decisions on satisfying (or "action to satisfy") actual needs, and for defining the economic, legal and environmental consequences

Medium-term planning - basis for decisions concerning locating and working out quantifiable needs and the analysis of the economic, legal and environmental consequences

Long-term planning - basis for reservation of resources or for demands for programmes concerning space and area of activity

In the short term, land utilisation and project planning can be unequivocally decided on. Long-term planning should take into consideration different development alternatives so as to be prepared for changed conditions and values.

Even if planning is carried out with a high degree of flexibility, new needs and demands can appear that are not in accordance with the plans worked out. Planning of this kind must be checked with regard to the new conditions of the project and the goals previously set up.

The real world we live in, and for which planning is carried out, is extremely changeable. With this background, modern planning should contain a built-in, continuous review of long-term plans.

4.3 Planning is carried out with differing degrees of geographical detail

Another type of classification refers to geographical extent, time limits and, as a result, degrees of accuracy.

Detail planning

Covers implementation of projects, parts of systems or urban areas, is detailed and concerned with short-term planning, possibly medium-term also, depending on the nature of the area.

Partial planning

Covers part of a community or an area limited by particular functions and is primarily general in nature. Concerned with medium- and short-term planning, but can also be concerned with long-term planning with regard to special reconstruction areas.

Overall planning

Covers a whole community or region (often part of a local government area). Concerned with long- and medium-term demands but also states short-term demands.

4.4 Participation of the general public

There have been strongly expressed demands from the general public for increased participation in the planning of their working, living and recreational environments. The principles governing the participation of the general public which are outlined by the Commission into Building Laws are being increasingly put into practice in current planning.

Participation in planning should be made possible at a sufficiently early stage, before too extensive preliminary investigations and design work take place, i.e. during the long- and medium-term planning stages. For example, it is essential to provide people with information including an account of the consequences of the current proposals for a subsurface project.

The following can serve as an exemple:
It happens relatively frequently that people living in the neighbourhood are totally unaware of the subsurface construction going on. It is only when damage occurs that the underlying cause is established, and people living nearby can suffer damage through no fault of their own. The forms for the provision of information and for the participation of the general public in the planning process, which are beginning to be developed, cover the needs as far as subsurface planning is concerned. In this connection it is assumed, however, that subsurface projects and planning are subject to the procedures referred to in building legislation.

4.5 Physical planning today

At present, planning is carried out partly with the support of building legislation and partly with the support of other laws.

Pending the new building legislation, planning land utilisation within communities is carried out according to the Building Law and supported by recommendations from the National Board of Urban Planning.

These recommendations incorporate concepts such as:

- broad general planning (master plan) for the community
- a plan for the disposition of land within the community as a whole (land use plan)
- partial planning (overall plan for a section of the local government area)

The practical control of both land utilisation and construction today rests partly in the Building Law and the building statutes, and partly in the regulations contained in the Law for the Protection of the Environment, the Nature Conservancy Law, road legislation, water legislation etc. A new method for use in planning within the community, known as the overview of local regulations (kommunöversikt), has been introduced in order to provide basic quidance for the control of construction outside the area of detail planning. The overview summarises all the regulations affecting construction, outlines current planning needs and lays down general principles for the examination of applications for building permission outside the area of detail planning. The overview of local regulations is provisional in nature but is seen as forming part of a broader master plan or land use plan.

Planning land utilisation is the co-ordinated planning of land as a resource. Another type of co-ordinated planning carried out within the area of responsibility of local authorities is long-term economic planning (KELP), where the limits on economic resources are stated.

Co-ordinated planning involving a statement of limits on resources also takes place in the planning of natural resources and energy and in social planning, to the extent that these occur. The Association of Local Authorities recommends local authorities to compile a register of general planning conditions which summarises the individual local authority's conditions and aims with regard to population, employment possibilities, construction of housing, physical planning etc.

4.6 Current revision of the Building Law

The 1968 Commission into Building Laws gives certain guidelines for social planning, as yet not permanently fixed, which influence the planning of today.

Experience gained from the current Building Law, basic principles of which have been in force since 1947, has led to a demand for radical reforms. The terms of reference of the work done since 1968 to reform the law and the Commission's results published in the spring of 1974 (SOU 1974:21 Land Utilisation and Construction - Principles for Legislation) have already made their mark on practical planning.

The aim of the work done on reforming the law is to produce legislation which partly co-ordinates all the regulations concerning land utilisation into a Land Utilisation Law, partly co-ordinates the regulations concerning construction into a Building Law, and adds to these a Law Concerning Local Authority Roads, etc.

However, the reformed building legislation will not be regarded as making redundant special laws such as, for example, the Law for the Protection of the Environment, the Nature Conservancy Law, water or mining legislation. In order that co-ordination can take place, however, it is suggested that the reviewing authorities regularly check that decisions taken do not conflict with the plan for land utilisation concerned.

In a report on the principles for a new Land Utilisation Law, the following new planning system is presented:

Master plan (overall plan for land utilisation within the community as a whole)

Partial plan (overall plan for part of a community according to need)

Detail plan (detail plan for implementation)

In general, it is assumed that the confirmation of a plan has a limited period of validity.

The principle report concerning the Land Utilisation Law does not take up the regional plan as a system of planning. Instead, when discussing planning between different local authorities, it assumes that a local federation will be formed in order to co-ordinate and direct planning for individual communities.

The report, which is very extensive, touches only very peripherally on subsurface construction.

It is stated that subsurface construction can cause ground-water drawdown, which can lead to considerable damage to existing buildings. The Commission considered that many improvements could be made in this respect by means of suitable regulations concerning planning and technical supervision in connection with subsurface construction and foundation work. The Commission therefore arranged to chart the legislative needs with reference to subsurface construction, ground surface works, ground-water and settling problems. The chart is probably the first of its kind.

The Commission into Building Laws assumes that land utilisation plans also cover subsurface installations and that subsurface construction will be subject to examination for building permits to a greater extent than at present. The Commission further assumes that the State and the County Councils, which are at present exempted from having to apply for building permission, will in future be obliged to do so. For reasons of security, defence installations are dealt with separately.

4.7 Subsurface construction in planning procedure

Subsurface construction can be taken into consideration both in co-ordinated planning by local authorities and in planning of particular functions.

Of special interest with regard to subsurface construction is planning of particular functions, which is dependent either on functional needs or on special legislation for particular functions. Some examples of the latter are plans for roads, plans for the supply of electricity, heating and water, and for sewerage etc., planning the construction of power plants etc. As mentioned by way of introduction, planning of particular functions requires special treatment and makes demands for large resources, in contrast to overall planning by society, which is co-ordinated and based on limited resources.

In <u>planning of particular functions</u> clear demands for subsurface space are thus expressed. These demands are mostly concerned with transportation, supply, storage, defence and industrial activities. This type of planning should be carried out with the same demands for time limits as in the co-ordinated planning of the local authority and should be able to express the needs, requirements and probabilities connected with subsurface demands.

<u>Co-ordinated planning</u> within the area of responsibility of local authorities, particularly the planning of land utilisation, which aims at anticipating all demands for land and the consequences of these demands, should also reflect the needs of subsurface construction. It should be the task of co-ordinated planning to provide necessary information, to co-ordinate surface and subsurface exploitation, to encourage the combined development of subsurface caverns and to make sure that conflicting demands do not prevent the development of the subsurface. If the demand for the examination of building permits is carried through with regard to the installation of subsurface caverns, in accordance with the intentions of the Commission into Building Laws, then the acceptability of subsurface construction and the integration of this into the urban organisation can be put to the test.

5. REVIEW OF INSTITUTIONS AND LAWS RELEVANT TO THE SUBSURFACE SECTOR

5.1 Activities in the subsurface sector at various institutions

5.1.1 Higher education

Traditionally, education within the subsurface sector taken place within the Departments of Mining and Metallurgy and Civil Engineering at the Institutes of Technology, while geology and hydrogeology are covered at universities. Two recently initiated courses of study are of particular interest:

- Courses of study in regional and urban development in the Departments of Architecture, Surveying and Civil Engineering at the Institutes of Technology have existed since 1971 (first graduates).

  Traditional teaching is combined and partly integrated with different subjects concerned with planning, town planning legislation etc.
- A course in soil mechanics has been established at the University of Luleå, with a special bias towards rock and installation technology. There is a special course in labour science.

5.1.2 State schemes

The Swedish Geotechnical Institute (SGI)

The Swedish Geotechnical Institute (SGI) has the responsibility of carrying out research and development work and, through its consulting activities, of providing services over the whole field of geotechnics. It also has the task of spreading information in geotechnical matters to the general public, decision-makers and specialists.

Its geotechnical research and consulting activities cover rock technology, geo-hydrology, physical community planning and use of automatic data processing methods. The Institute has also been given the task of monitoring soil movements and stability conditions in areas where there is danger of subsidence. Its information work comprises the maintenance of a survey of domestic and foreign literature, the publication and distribution of articles and reports, together with the running of a central geotechnical library.

The Institute is situated in Linköping and has about 70 employees. Its activities are spread over three research and consulting departments, an information unit, laboratories, field work and measurement section, and a store/workshop.

The primary duties of the department are in this connection:

- general geological surveying
- geological documentation and information (ground water, geophysics)
- work on assignments (ground-water surveys, geological construction surveys etc.)

In 1978, SGU was relocated from Stockholm to Luleå. The staff numbered 530 in 1975 and was distributed as follows: the bedrock division - 202; the quartenary and hydrogeological division-83; the geophysical division-115; the geochemical division-60; and the administrative division-70.

The Geological Survey of Sweden (SGU)

The aims of work carried out by the State in the field of geology are:

- to prepare a basis for the planning of the use of terrain, particularly with reference to bedrock, soil layers and ground water
- to assist governmental and local authorities, companies, organisations and individuals by providing expertise and services within the field of soil mechanics.

The Swedish Office of Industrial Safety

The Swedish Office of Industrial Safety, i.e. the Swedish Board of Industrial Safety and the Labour Inspectorate, are to exercise supervision over the application of the Labour Welfare Act and statutes subsequently established, i.e. over all questions concerning the working environment. The Swedish Board of Industrial Safety gives notice of regulations with regard to protection against industrial injury and also carries on object and basic research in questions concerning the working environment.

In accordance with the Bill on Working Environments (SOU 1976:1), the Swedish Board of Industrial Safety is to provide sanction-backed regulations on working methods and installations, machines and other technical devices and their use, amongst other things under the surface. Within the Swedish Board of Industrial Safety questions concerning the subsurface are dealt with by the Mining Office within the Building Section of the Supervision Department.

The General Labour Inspectorate is organised over 19 districts each covering one, but in some cases, two counties. It is their duty, which involves work in the field, to see that regulations regarding safety and working hours are followed and to give advice in questions concerning the environment.

5.1.3 Research

Rock Mechanics research is to a considerable extent carried on by

- The Swedish Mining Association, principally within the fields of:
  - . geology and prospecting rock technology
  - . mineral technology
  - . mining machine technology
  - . industrial safety
- The Swedish Foundation for Research in Detonics principally within the fields of:
  - . detonics
  - . environmental technology
  - . rock mechanics
  - . mining technology
- The Swedish Rock Mechanics Research Foundation, BeFo principally, within the fields of:
  - . preliminary investigations
  - . safety and environmental protection
  - . design methods
  - . rock disintegration
  - . mining methods
  - . rock reinforcement
  - . special installations

The above mentioned research is carried on in close co-operation both between the three organisations and with the Institutes of Technology, authorities etc.

The Swedish Academy of Engineering Sciences' Committee into Subsurface Construction is involved in and works for creating the necessary initiative in questions not directly related to rock mechanics.

The Rock Blasting Committee is an independent group supported by subscriptions from companies whose work involves facilitating the exchange of experience within the field of mining.

The Society for Building Geology is a trade association with the aim of promoting building geology in Sweden. Its work mainly comprises information and study days.

Rock Mechanics Research within governmental departments is in some cases extensive, for example, at the National Swedish Power Administration, the Swedish Fortifications Administration, the Geological Survey of Sweden, Swedish Geotechnical Institute - SGI, the National Road and Traffic Institute etc.

Research within companies is considerable. Mining companies and manufacturers of various types have long been active. State research funds set up at the Technical Development Board, the Swedish Council for Building Research, and the Fund for Industrial Safety, have meant that consulting and contracting companies have increased their research activities.

Research at the institutes of Technology and Universities

This has to date been to a great extent limited to the Department of Rock Technology at the Royal Institute of Technology, and to a certain extent to the Departments of Agricultural Hydrotechnics and Civil Engineering. It is intended that research be carried out at the recently created Department of Soil and Rock Mechanics.

Considerable resources have been made available at the new University of Luleå in the Departments of Rock Technology and Rock Mechanics.

Research is also carried out at

Chalmers University of Technology:
the Departments of Soil Mechanics, Geology and Hydraulic Engineering

University of Luleå:
the Department of Geology

At Uppsala University research is being carried out into geological and seismological engineering questions.

State-financed rock mechanics research

The comprehensive research work of the last few years has partly and sometimes totally been financed by the state through the Technical Development Board (STU) and, since 1971, through the Swedish Council for Building Research (BFR). The newly set up Fund for Industrial Safety has also, through its support for different environmental projects, facilitated research activity.

5.2 Review of current laws and regulations concerning the subsurface

There are no special laws concerning subsurface construction in general. The principal in existing legislation is that the kind of activity is not crucial; legislation is concerned only with intended use. We have very carefully regularised legislation concerning wastewater, legislation which has not, however, been drafted for subsurface construction but for the discharge of wastewater. We have, to take other examples, regulations concerning storage of inflammable goods, and, similarly, there is a law concerning electrical installations. These regulations are valid regardless of whether the installation is above surface or whether it is subsurface in the form of a tunnel or rock chamber. We have attempted to summarise existing legislation in the diagram on the next page.

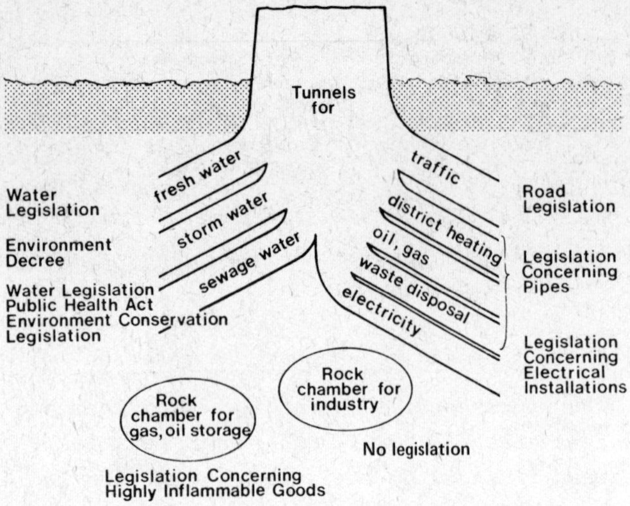

General legislation concerning subsurface construction is characterised by the headings above the stump. The various roots are here meant to represent various types of tunnel and cavity. There are then special laws and regulations which may apply in the case of the respective tunnel installation according to the use to which it is put. For several types of tunnel there is no special legislation, in which cases we have put a question mark.

5.2.1 Planning and design stages

It is necessary in the planning and design of fixed installations to take into consideration the numerous laws and regulations which have come into force at different times during the present century. The result of this can best be described as a patchwork made up of patches of every conceivable size, shape and colour. The patches do not of course always fit; sometimes there are gaps and sometimes there is overlapping. The existing legislation should thus with some justification be characterised as difficult to review.

The Building Law

The first paragraph of the Building Law is as follows:

> "The use of land for building on shall to the extent prescribed in this law be preceded by planning. Permission is required for building (building permission) in the scope decided by the government".

The wording does not exclude subsurface installations, but has in accordance with current practice been restricted to apply only to construction work wholly or partly above ground. The law cannot thus form the basis for a requirement that subsurface construction shall be covered by planning or controlled by building permission.

The following planning instruments are taken up in the Building Law:

Regional plan
General plan, (for determination, or only advisory, without statutory force)
Detail plan, town plan or building plan

It is generally assumed that building is regulated by planning. Building outside the scope of the plan is considered in connection with application for building permission.

The Building Code

Building permission, which is dealt with in the Building Code, is probably not required for most subsurface installations, but is necessary as soon as an installation breaks the surface. Building permission may contain more or less detailed stipulations concerning the construction stage.

Water legislation

Where construction work affects lakes, rivers, ground water and the like, it may be the case that permission in accordance with the Water Law has to be applied for.

The Environment Protection Law and Act

If the installation is subsequently used for purposes which can according to the Environment Protection Law or Environment Protection Act be regarded as activity detrimental to the environment, permission or dispensation for operation is required. This is granted by the Concessions Board, the Crown or the County Administration Boards. The required permission must be applied for at the planning stage so that design planning can be adapted to follow those conditions which are required in the concession or dispensation which may be given.

Nature Conservancy Law

The Nature Conservancy Law can apply if the external environment is affected, in the case of

- areas or sites which are considered valuable from the point of view of nature conservancy
- depositing of surplus excavated material. It is prescribed in the law that certain enterprises, which can change the natural environment in a fundamental way, shall be the object of discussion with the County Administration Boards before such enterprise is embarked on.

5.2.2 Construction stage

The Building Law

There are no regulations which give the Building Committee the responsibility of checking the site for an installation, information on the final location and layout of the rock cavity, its imperviousness or its hability to water leakage and its consequent effect on the environment.

The Industrial Safety Law

The Industrial Safety Law includes general regulations concerning occupational health for the majority of occupations, and thus includes work subsurface. Since the main content of the most important regulations are probably well known, no further details of the law are given.

The Industrial Safety Act

The Industrial Safety Act makes up a collection of directives concerning the application of the Industrial Safety Law. The directives concern all types of work, and thus work both above and below surface. It is, however, expressly prescribed that necessary ventilation for work underground be arranged (20 §). In addition, the National Board of Industrial Safety has issued special directives concering work underground, i.e. those referred to as the "rock directives."

5.2.3 Utilisation stage or operation period

As in the case of the construction stage, there is no responsibility on the part of the Building Commission to carry out control of the condition of the rock cavity or of its imperviousness to inflow of water.

Legislation as regards permission, concession, dispensation, application etc. has been collected together. This has been done to ensure acceptable use of an installation both from the point of view of the individual employee and the surrounding area. In many cases these decisions include regulations concerning operation (of an installation) and during operation time supervision is carried out to ensure that the regulations and conditions which are associated with the permission are followed by management. Industrial safety legislation includes in addition detailed regulations concerning the internal environment of a subsurface installation. The possibilities of intervention during the period of operation are given in the Nature Conservancy Law, among other places.

Responsibility for ensuring that the regulations associated with concessions in accordance with the Nature Conservancy Law are followed rests partly with the Nature Conservancy Department, which is responsible for central supervision, and partly with the Public Health Boards and the County Administration Boards, which are responsible for the day-to-day control.

The Labour Inspectorate, the Swedish Board of Industrial Safety and the Public Health Boards ensure that the Industrial Safety Law and the Industrial Safety Act are followed. It is primarily the Building Committee but also sometimes the Public Health Board which are to ensure that the Building and Public Health Codes are followed during the construction and operation stages. The latter is true in the case of Stockholm.

5.2.4 Right of disposition

The right of disposition for land which is owned by another party to be used for subsurface installations is normally acquired through an easement agreement or usufructuary agreement with the owner of the land. These forms have certain limitations - an easement agreement can only be reached between property owners and a usufructuary agreement gives right of access for only a limited period of time. Right of disposition can also be acquired through compulsory powers, sometimes with preceding agreement (pro forma), which usually in that case refers to right of usufruct for an unlimited period.

A subtler form is applied in legislation (1974) concerned with roads and pipelines, where an installation in its entirety is investigated by an executive officer. (In the forms referred to above, where several properties are concerned, a separate agreement must be reached for each property). The latter law applies in the case of

- telecommunications
- high voltage current
- water
- sewage
- district heating, oil, gas
- purposes for which concession has been obtained

The interpretation of this legislation in questions concerning the subsurface is not entirely clear, as the law has not yet received wide application.

5.2.5 Responsibility and damages

Requirement for proof of negligence - connection between activity and damage - burden of proof

In principle, damages shall be awarded to a claimant for damage or negligence caused by another party. According to case law, strict responsibility applies if works are undertaken with great risk of damage to the surroundings as a result of, for example, the unusually great scope of the works. Subsurface works are probably in most cases to be counted as such.

For the question of responsibility to arise it is necessary that the damage is a measurable result of the construction activities. The burden of proof rests with the party suing for damages, i.e. the claimant must show a connection between the activity and the damage. (Something which in these context often proves to be a difficult and costly task.)

Which laws can be involved in case of damage?

The position in law before 1969 was largely determined by certain leading cases in the High Courth (the law of precedent), which made up case law. In line with development, the attitude of the High Court also changed, the result of which was a kind of set of fluid regulations which could be rather difficult to understand for the layman. During the last few years, however, case law has in part been summarised in some new "weighty" laws, namely

- the Land Law, 1972
- the Law of Damages, 1972
- the Environment Protection Law, 1969

The <u>Land Law</u> deals with, amongst other things, damage resulting from excavation, piledriving, laying of sheet-piles or blasting, in which cases strict responsibility is normally restricted to works extending below normal cellar depth. Changes to the ground water could probably come under this law (under the section on water, and also under the Water Law). Responsibility rests with the property owner.

The Law of Damages complements the Land Law and the Environmental Protection Law. Damages may be awarded for injury or damage to persons or property. The law places responsibility on the person causing the injury or damage (or on this person's employer), regardless of status (proprietor, contractor, possessor of the right of usufruct, etc.).

In the Environment Protection Law, provision is made for responsibility in the utilisation of land and for works which are not of a wholly temporary nature. The law does not apply in cases of damage arising from drawdown of the ground water. The concept of strict responsibility according to the Environment Protection Law goes further than the regulations on responsibility in the Land Law. It is unclear whether the law applies only to the property owner or whether it can be made to apply also to a contractor.

The new laws clearly encroach on each other in several respects and compete with each other, which is why it is very unclear where the boundaries of the areas of application of the different laws lie. The situation is further complicated by the fact that the Land Law and the Law of Damages are handled by ordinary courts of law, while cases concerning the Environment Protection Law are dealt with by a special court (Fastighetsdomstolen). It is thus important right at the start to decide which law one wishes to invoke at a court case since a demand for damages must be based on either, for example, the Land Law or the Environment Protection Law, not both, in one and the same suit.

5.2.6 Official Secrets Legislation

The question of secrecy as regards subsurface installations may arise in two contexts: for reasons of defence and in investigations of a so-called internal nature.

In the first case, the Official Secrets Law is supplemented by directives and regulations in the Security Act and in a special statute on maps (kartsekretessförordningen).

Every case concerning secrecy must be able to be supported by an explicit regulation in the Official Secrets Law. Important in this respect is that it is the kind of activity which is secret and not, for example, the subsurface installation itself. The question of whether a document is secret is dealt with by the authority having safekeeping of the document. Special rules apply in the case of documents of a partially secret nature.

The bringing under the official Secrets Law of investigations of a so-called "internal" nature is supported by § 9 of the Official Secrets Law. Here it is laid down that, amongst other things, investigations of a technical kind which an authority has had carried out on behalf of the general public can be held secret by this authority for up to 20 years.

INVESTIGATED TOWNS

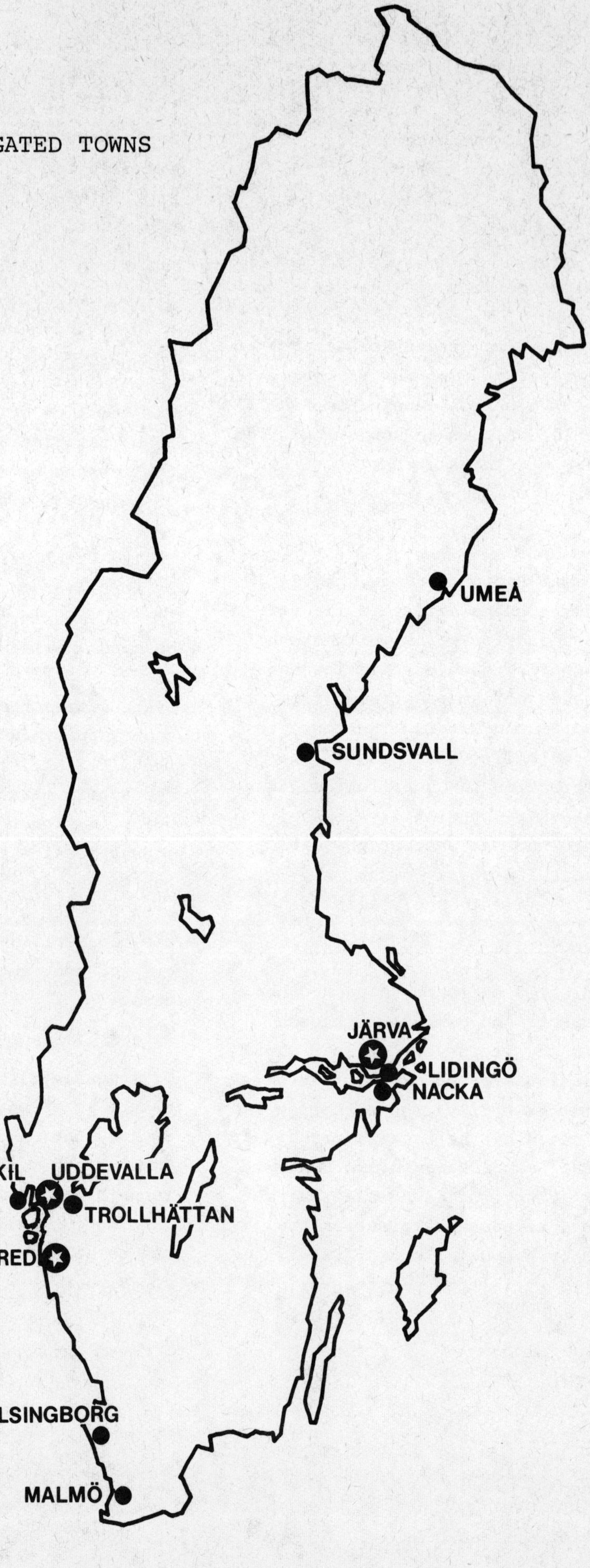

6. INVESTIGATIONS WITHIN THE PROJECT

In order to create basic material for appraisal of subsurface construction in Sweden and of the demands that should be made on planning and practical application, a number of surveys have been carried out within the framework of the research project. These surveys have been concerned with:

- a survey of the internal environment in subsurface installations
- the plotting of subsurface construction in eight communities
- analysis of subsurface construction in three areas
- descriptions and registration of the position and shape of subsurface installations
- financial evaluation of subsurface location
- permits and regulations, responsibility and compensation for damage
- ecology
- international experience

The three first-mentioned surveys are presented in more detail below.

A preliminary summary of experience acquired was made during the spring of 1975 and sent to about 100 public bodies in order to obtain their opinion. Answers were received from 60 of these. The distribution is shown on the table below:

Category of public body	Number
Government departments	17
Provincial authorities	4
Universities and equivalent institutions	10
Local authorities	12
Central associations, etc.	3
Contracting companies	5
Industrial companies	2
Consulting companies	5

The answers represent an important contribution to the research project both in the form of comments on the summary, and also in the form of new material.

Summary of subsurface installations in eight communities

An overall picture of the size, type and distribution of installation and the population per built-up area (1975) is given in the table below.

	Umeå	Sundsvall	Lidingö	Nacka	Trollhättan	Lysekil	Helsingborg	Malmö	Total
Population in built-up areas, 1,000 inhabitants	48	54	36	47	41	7	82	265	
No. of installations existent or under construction									
civil installations	3	14	19	27	18	4	15	101	201
defence installations	2	1	-	-	2	2	3	1	11
Volume of blasted rock, 1,000 m³	300	92 x)	384 x)	880 x)	360	1306	5	42	3369
Volume excluding oil storage installations	60	92	376	773	360	36	5	42	1744
Length of tunnels, km	3.7	0.04	14.8	17.7	28.0	2.1	0.03	1.9 x)	66.3
No. of installations planned or under consideration	1	-	≥5	3	≥3	large	3	≥10	≥25 excl. Lysekil
Distribution of installations									
sewage treatment plants	-	1	1	1	1	1	-	-	5
supply tunnels	1	1	14	12	12	1	-	1	42
oil storage installations	2	1	2	3	-	2	-	-	10
sub/transformer stations	-	6	-	-	-	-	-	77	83
sand silos	-	1	-	1	-	-	-	-	2
static water tanks (fire)	-	2	-	-	-	-	11	17	30
work shops	-	-	1	-	1	-	-	-	2
bomb shelters	-	-	1	6	1	-	-	-	8
communication tunnels	-	-	-	3	-	-	1	2	6
other installations	-	2	-	1	3	-	3	4	13

x) Information on volumes missing for 1 installation in both Sundsvall and Lidingö and 4 installations in Nacka. Information at length missing for two tunnels in Malmö.

Other installations include: water reservoirs, power stations, subsurface toilets, sewage pumping stations, telecommunication stations, control tunnels, test mines (compressed-air storage).

6.1 The internal environment in subsurface installations

An investigation, which is completely independent from the other surveys, has been carried out within the fields of industrial health and working environment.

The survey work started by isolating the problems and by making up an inventory of Swedish research efforts concerning the subsurface environment. It was found that research on the risk of damage during the construction stage is being carried out at many places in Sweden, as well as in the rest of the world, within industry and commerce, at universities and other institutes. Since the most serious problems occur during the actual construction stage, it would appear natural for research resources to be concentrated to this particular field.

On the other hand, there would only appear to be a few surveys concerned with psychological and physiological reactions to different components of the subsurface environment during the utilisation stage. An extensive survey of the literature showed that the range of knowledge in this respect is poor, even on an international scale.

The survey has also included extensive work on compiling information, partly from the various trades where surveys have been carried out which can to some extent be applied to time spent in subsurface environments, and partly from people who have been in contact with the problems. In connection with this work, about 80 interviews and surveys of the literature have been carried out at about a dozen libraries and data banks.

The result of the work is described in section 9.4.

6.2 Subsurface construction in eight communities

The aim of the survey has been to use random sample surveys in order to create a picture of current construction of subsurface installations from different perspectives. The material has been compiled through interviews, primarily with officials in local authorities.

In the choice of locations, which was carried out in joint consultation with such authorities as the Swedish Board of Urban Planning, it was assumed that there was a certain amount of subsurface construction. A geographical spread was desirable, with representatives for southern, central and northern Sweden. Places were chosen which had geological conditions other than primary rock (Scania). Communities within city areas, and thus with regional relationships, were to be included, as well as communities out in the country. The aim was to underline differences related to the size of the communities (Population).

Questions took up the following:

- existing and planned subsurface installations
- organisation, administration with respect to subsurface construction work
- experience of planning and design work
- experience of construction and operating work
- requirements and opinions on
  - . basic material for planning, planning methods
  - . arguments in favour of and against subsurface construction.
  - . legal, organisational and administrative aspects
  - . the financial aspect.

Comments on results

A summary of information on subsurface installations in the communities is shown in table

An attempt to summarize the different installations is shown below.

	No. of installations	Volume blasted m³ (1,000)	Tunnel length km
Oil storage installations	7	1685	-
Other installations	87	1640	-
tunnels	46	480	66.3
cavities	41	1160	-
Total	94	3325	66.3
Note: Henriksdal and Käppala sewage works	2	800	

The following limited assumptions have been applied in the summary:

1. Malmö has been excluded. The installations are small and not comparable with those in other communities.
2. Defence installations are not included.
3. There is no available information on six installations.

Comments

1. The volume of oil storage (78 % of which is in Lysekil) is of the same order as the combined total volume of all other installations.
2. Half the volume of other installations comprises two regional sewage treatment plants.
3. If the above mentioned large installations are excepted, the remaining installations comprise 46 tunnels and 39 rock chambers. If an average for each community is calculated then each has:

   7 tunnels, each 14 km in length, with an area of 7 m², and 6 rock chambers, each with a volume of 9,000 m³. This gives a cavity volume of 3 m³ per inhabitant distributed over the population of built-up areas. If, however, the large installations are included, then a figure of 190 m³ per inhabitant is obtained for Lysekil and an average of 9 m³ per inhabitant for the other communities.

Distribution in terms of year of construction is as follows:

Time period	-45	46-55	56-65	66-75	66-75x)
No.	12	19	28	35	31
Volume (1,000 m³)	160	250	420	2500	430

x) Installations constructed between 1966-75 except the oil storage installations in Lysekil and the Henriksdal and Käppala sewage treatment plants.

Motives for subsurface location

The motives given for the subsurface location of the different installations are summarised in the table below. The table is of an indicative character: in a single local authority several installations can have been prompted by one and the same factor and, conversely, an individual installation by several different motives.

Charts

Subsurface construction is illustrated in chart and table form for three of the eight communities. Lysekil represents a small built-up area with close connections to a refinery. The integration of Lidingö into the Stockholm region is apparent not least by its subsurface construction. Finally, Trollhättan is a typical medium-sized Swedish town markedly affected by overall local authority planning.

Motives for subsurface installation	Umeå	Sundsvall	Lidingö	Nacka	Trollhättan	Lysekil	Helsingborg	Malmö
Desirable/necessary as a result of restrictions on surface use	X	X	X	X	(X)	X	X	X
Environmental demands		X	X		X	X		
Civil defence aspects		X	X				X	X
Subsurface location economically beneficial		X	X	X	X	X		
Climatic protection (against cold)		X				X		
Nature conservancy demands			X					
Legal reasons (disposition of land)			X					
Future capacity			X					
Potential joint use	X		X					
Desired flexibility in surface use			X	X				
Technical reasons (gravity tunnels)				X	X			
Good bedrock				X	X			
Topographical obstacles				X		X		
Extreme case solution								X
Confidence in tunnel construction (tradition)					X			

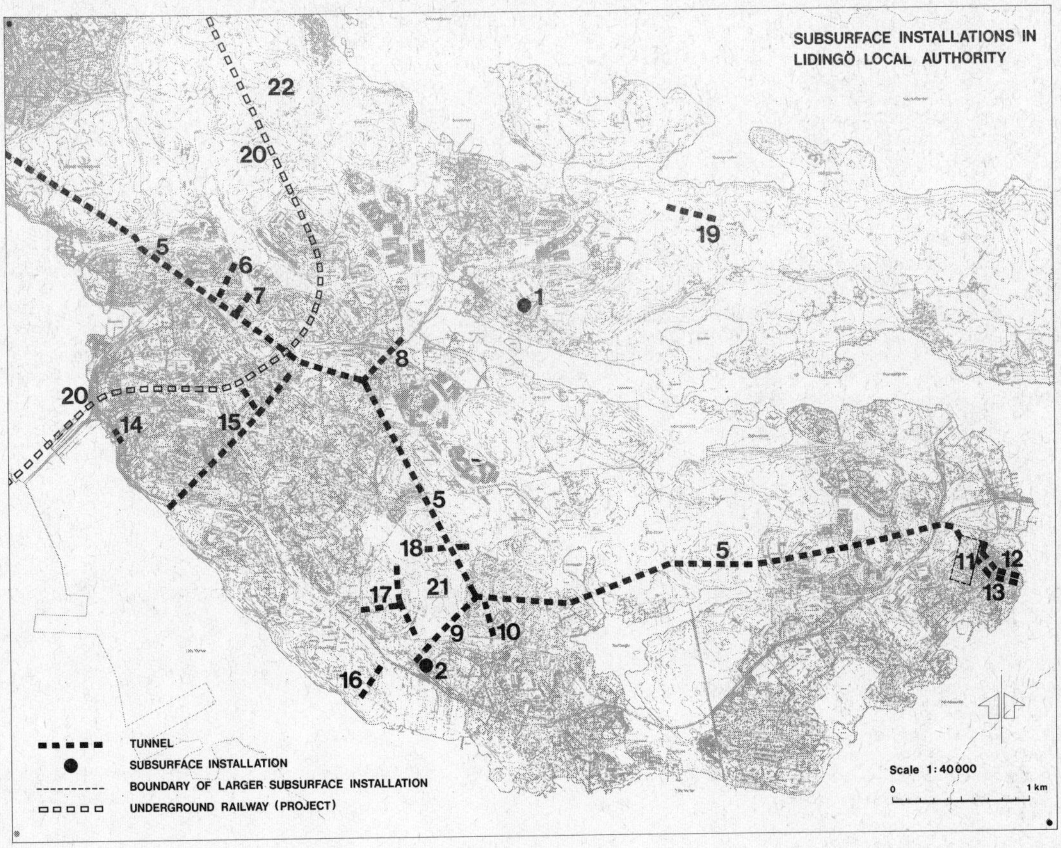

Number, on map	Installation	Construction time	Length m	Cross-section m²	Volume m³
1	Bomb shelter	1940	-	-	2600
2	Workshop	1940	-	-	-
(3-4)	Oil storage sewage	1945	-	-	-
5	Tunnel (total length 60 km)	1958-63	8900	8	71000
6-10	Connecting tunnels	1958-63	1800	-	30000
11	Käppala waste water treatment plant	1965-69	-	-	250000
12-13	Discharge tunnels	1959-63, 1965-69 resp.	500 400	4.5 11	2250 4400
14-19	Tunnels for storm water, some including pipes for waste- and fresh water	1968, 1972-74	3150	5	15750
Installations, under consideration or planned					
11	Extension of the Käppala plant	1977-79	-	-	85000
20	Underground railway	-	-	-	-
21	Further storm water tunnels	-	-	-	-
22	Possible future storm water tunnels below planned town area	-	-	-	-

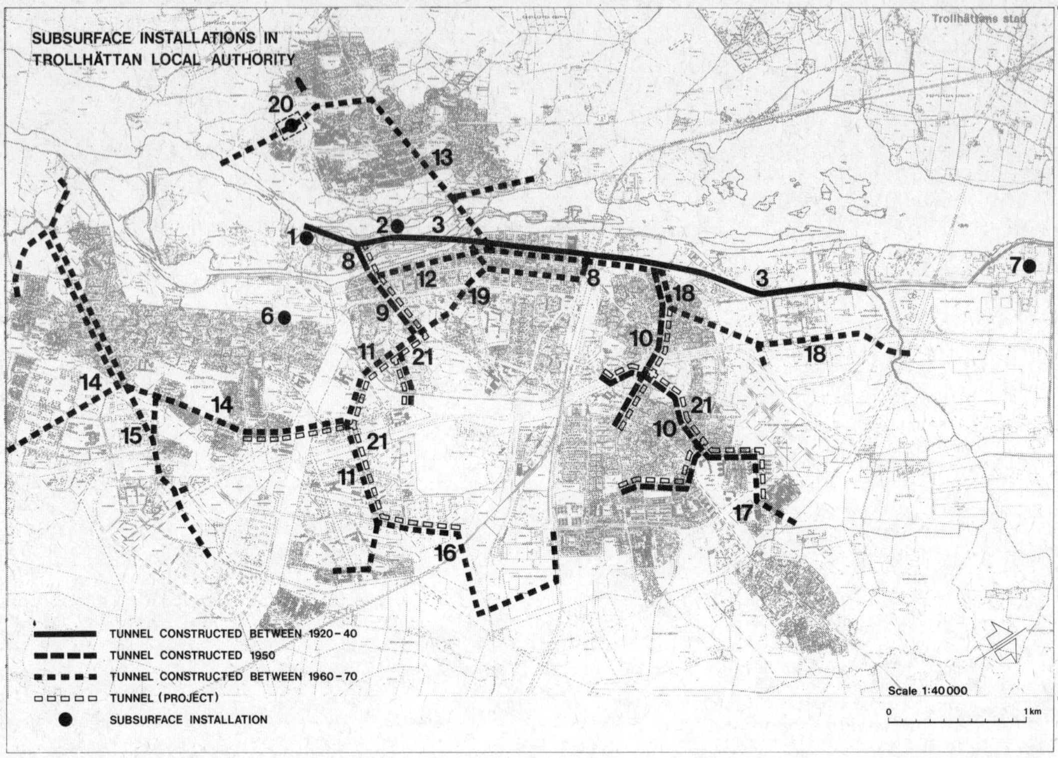

Number, on map	Installation	Construction time	Length m	Cross-section m²	Volume m³
1	Hydro-electric power plant station	1909-18	-	-	19000
2	" " "	1942	-	-	60000
3	Tunnel for regulation of water level	1925	3300	4-8	28500
4	Bomb shelter	1955	-	-	-
5	Civil defence installation	1955	-	-	-
6	Tunnel for factory	1955	150	20	3000
7	Workshop	1952-53	-	-	100000
	Compressed air storage	1953	-	-	10900
8-18	Sewage and storm water tunnels	1930 to 74	23000	3-8	100000
19	Sewage and storm water tunnel	1974-76	1600	8	12800
20	Waste water treatment plant	1965-67	-	-	25000
Installations, under consideration or planned					
21	Future system of separated storm water tunnels	?	7500	8	60000
20	Extension of waste water treatment plant	?	-	-	25000

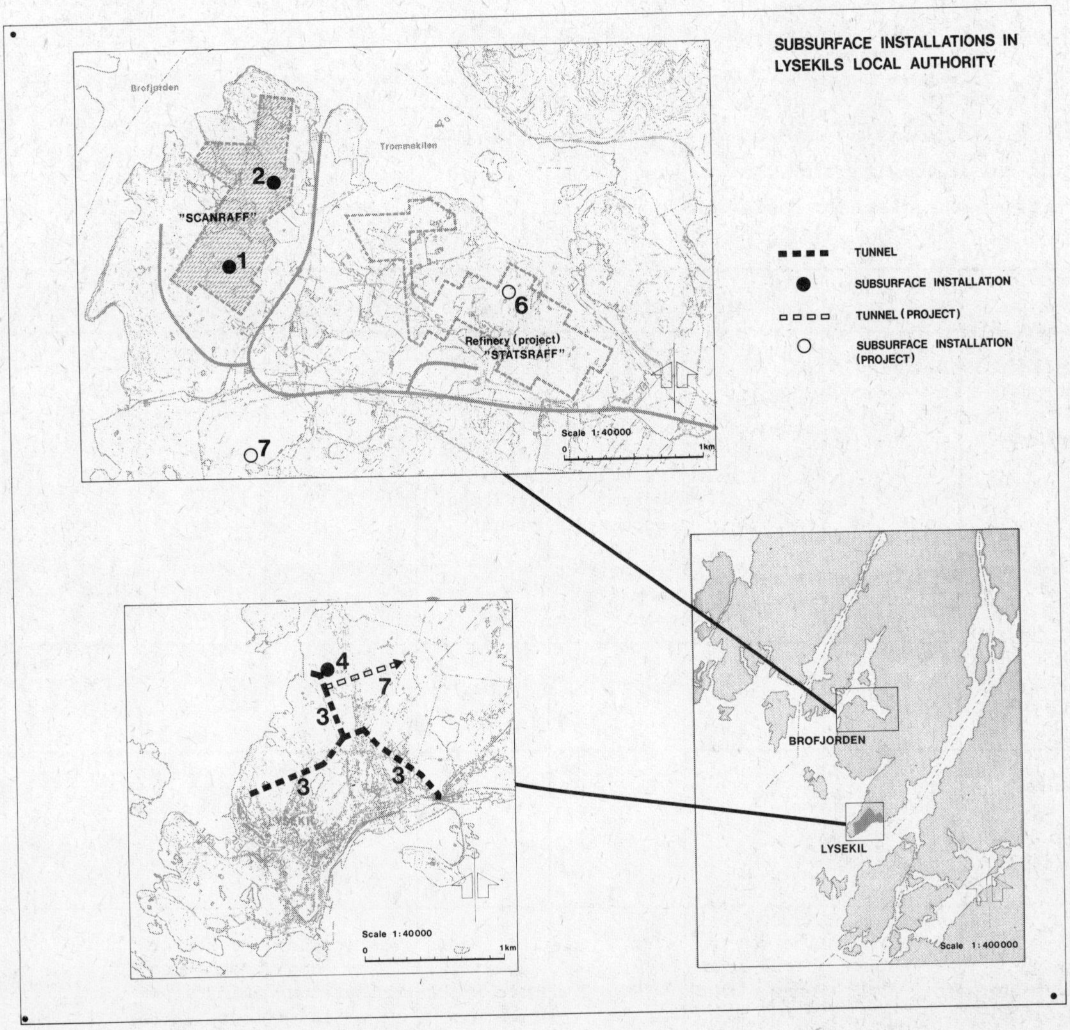

Number, on map	Installation	Construction time	Length m	Cross-section m^2	Volume m^3
	Oil storage for SCANRAFF				
1	4 cavities for storage of crude oil	-	-	-	very large
2	4 cavities for storage of refined products	-	-	-	"
3	Sewage tunnels	1974-76	2100	5	10500
4	Waste water treatment plant	1974-76	-	-	25000
(5)	Civil defence installation	-	-	-	-
Installations, under consideration or planned					
1	Plans for extensions	-	-	-	-
6	Plans for oil storage	-	-	-	-
-	"	1976-81 ?	-	-	large
-	Possible power station	-	-	-	
7	Planned sewage tunnel				

Planning, administrative management, experience, aims

A general summary of the work methods and experience of the local authorities presented in question and answer form is given in the table below. The summary is of interest from the planning point of view.

Question	Umeå	Sundsvall	Lidingö	Nacka	Trollhättan	Lysekil	Helsingborg	Malmö
Are subsurface installations included in the overall plants?	-	All installations are parts of a larger supply plan	The Käppala plant is a regional sewage treatment system. Technical supply is prescribed in master plan	The Henriksdal plant is a regional sewage treatment plant	-	-	-	-
Are tunnels put to multipurpose use?	Yes	Not relevant	Yes	No	No	Not relevant	Not relevant	-
Have problems of crowding been observed at the design stage?	No	No	No	Yes	Yes	No	-	-
Have cost-benefit analyses been used when choosing different subsurface alternatives?	-	The reason "good economic situation" is given	Yes	The reason "good economic situation" is given	The reason "good economic situation" is given	The reason "good economic situation" is given	-	No
Which department of the local authority deals with questions of subsurface planning and design?	Public Works Dept.	Public Works Dept. Electrical Dept.	-	Public Works Dept. Henriksdal plan - part of Stockholm local authority	Public Works Dept.	Public Works Dept.	Public Works Dept.	Public Works Dept.
Has building permission been applied for?	-	No	Only for certain surface installations	-	No, storage installations	Yes, for oil - not local auth. installations	Sometimes	In certain cases
Are subsurface installations included in the town plan?	No	No	Yes	-	No	"	No	Sometimes
How has right of disposition been obtained for subsurface installations?	State-owned land	Local auth. owned land	Easement agreement, a few cases of compulsory purchase	Agreement with land owner	Civil law agreement. No easement agreement. No compulsory purchase. Tunnel moved if permission refused	Easement agreement. Tunnel moved if permission refused	-	-

Question	Umeå	Sundsvall	Lidingö	Nacka	Trollhättan	Lysekil	Helsingborg	Malmö
How is handling of excavated surplus material organised?	Extension of port installations	-	-	-	Used in road construction	Special plan for handling of excavated material. Working group for co-ordination	-	-
Are subsurface installations registered in the same place?	-	No	Partially	-	No	Yes	No (very poor knowledge of 2 installations)	No
Has the volume of completed cavities been measured?	-	-	-	Partially	No	Yes, as well as measures for rock reinforcement	No	-
How is less current geological info. used?	-	-	Divided archives inventory through individual memory	-	-	Previous investigations found by reference to inventory	-	-
Have deficiencies in the scope of previous investigations been observed?	No	Yes	No	-	Yes	No	-	-
What types of damage have been noted?	None	Rock falls, blasting damage, large-scale consequences	Some damage to wells and buildings	Falling rocks in water supply and sewerage tunnels. Dry wells	Damage due to vibration. Groundwater situation unclear	-	-	Dry walls
What information needs do the local authorities have?	Summaries of laws and regulations. Rules for responsibility and settlement of claims	Settlement of claims	None	-	-	-	-	Subsurface experience registered as small
How should subsurface installations be dealt with in overall planning?	Must be integrated in surface planning	-	-	-	-	Surface and subsurface mutually dependent. Co-ordination difficult but necessary. Co-ordination with surface town plan desirable	-	-
In other planning?	Cost-benefit analysis necessary	Geohydrological info. needed	-	-	-	-	-	-

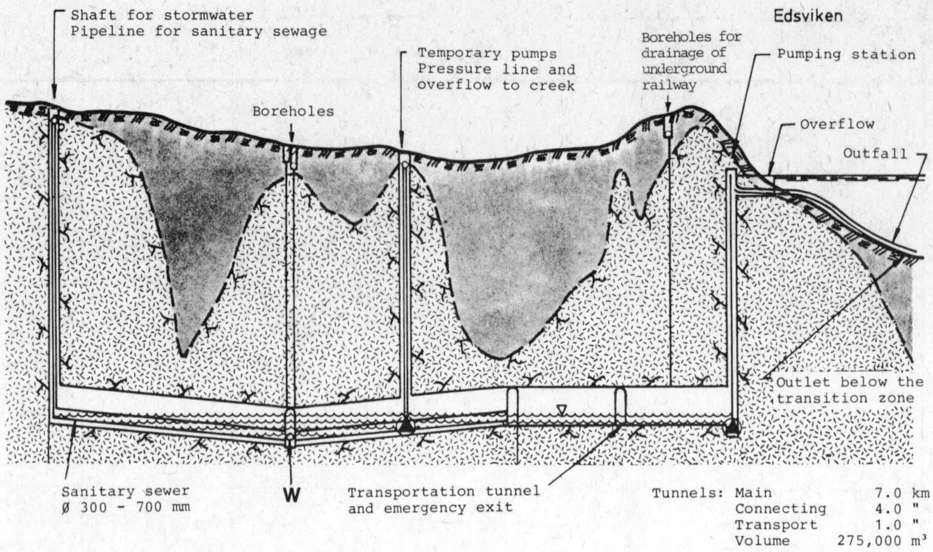

Figure 9

Combined storm- and wastewater tunnel at Järva. The storm water tunnel is actually a reservoir which is sufficiently large to contain the collected storm water which is then pumped out into the Baltic (Edsviken). In parts of the tunnel here, the wastewater is led into pipes under pressure towards a separate wastewater tunnel (position W).

Stockholms VA-Verk

Figure 10

Pre-sedimentation tank at Käppala sewage treatment plant, Lidingö. Location in rock provides operational advantages primarily because of climatic protection against high and low temperatures.

Vattenbyggnadsbyrån

6.3 Survey of three areas

In a larger study, detailed analysis has been carried out of all subsurface installations in three areas. In the case of each project, the entire process of execution has been plotted from the preliminary idea stage to the completed installation, as well as the relationship to other planning activities. Two of the areas selected, Angered (north-east of Gothenburg) and Kista-Husby-Åkalla, Järvafältet (north of Stockholm) are examples of new use on a large scale. The third, the centre of Uddevalla, is an example of an older town which has been affected by extensive works involving new installations. In these three areas, several subsurface installations are being built or have been completed.

The work has been carried out primarily through interviews with a very large number of people within various local authorities, at consulting companies, contracting companies and at places of work. Fairly complete material has been compiled in the form of the current planning instrument, surveys, drawings, tender documents, geo-surveys, etc. It has been found relatively easy to discover material with effect from the time at which decisions have been made concerning the installations, that is to say, from the design stage. It was more difficult to find documents relating to planning considerations. The reason for this is that such considerations have often taken the form of meetings and conferences and have only been recorded to an extremely small extent. The extensive periods of time concerned in many cases - from planning to completion, often more than five years - has naturally complicated the possibility of reconstruction.

Angered-Järva

Angered (north-east of Gothenburg) and Kista-Husby-Akalla (Järvafältet in Stockholm) are both urban areas, the construction of which started in the 1960's, and which are still being developed. Population figures are expected to reach between 30,000 and 40,000, and both areas include a number of industrial areas. Both areas have tunnels for rail traffic and tunnel systems for supply.

The planning of the areas, because it is on such a large scale, is of a kind which is seldom found. Development occurs within a short limited period of time and the completed community neither permits nor is considered to require any larger degree of flexibility. It is therefore more a question of implementation planning.

Public transport is based on a rapid tramway system in the first case and on a metro system in the latter and comprises an integrated element in the construction of the communities. The transport elements have influenced planning more than any other functions. The surface and subsurface have thereby been securely bound together.

Supply from a technical point of view has been constructed on the basis of completely different criteria. The relatively general picture of the planned construction has provided the starting point for the determination of the principles of supply. The tunnels have then been designed to provide the best stretches between starting and end points from a technical and economic point of view.

Uddevalla

Subsurface construction in Uddevalla was initiated just after the turn of the century. The construction of the railway involved the tracks passing through six tunnels. This was because of the hilly terrain.

During the war, defence installations and the Fossumsberg hydro-electric power station were constructed. During the fifties, further defence and civil defence installations were constructed. Local authority subsurface construction began at the end of the fifties with some smaller tunnels for

Summary of subsurface installations in three areas

	Length of tunnels km	Volume of blasted rock (1000 m³)	Completed in
Angered			
Supply tunnels			1972
Multi-purpose tunnels	8.5	187	
Wastewater tunnels	8.2	49	
Storm water tunnels	4.4	80	
Transportation tunnels	2.0	37	
Tramway tunnels including subsurface station	2.0	110	1971
Total	25.1	463	1971
Järva			
Supply tunnels (technical)	7.2	100	1971
Storm- and wastewater tunnels	13.0	250	1977
Underground railway, including 2 subsurface stations	1.9	140	1977
Total	22.1	490	1977
Uddevalla			
9 railway tunnels (6 old, 3 new)	1.9	71	1905 1970, 75
2 motorway tunnels	0.6	72	1976
11 pipe/cable network tunnels	3.3	33	1957, 75
Power station	-	3	1943
Sewage treatment plant	-	105	1974
Bomb shelters	-	8	1962
Sand silos	-	2.4	1959
Defence installations (4)	-	-	1943, 58
Total	5.8	294	

In addition there are in Uddevalla some 10 installations which are the subject of planning or are under consideration.

water supply and sewerage. Two years later, a bomb shelter for the Uddevalla shipyard was constructed. A through-road, which has recently been completed, has resulted in the construction of a whole series of subsurface installations. The motorway itself passes through two hills by means of tunnels. The direction of the road has forced realignment of the railway, which in three cases passes through tunnels. In connection with the coordination of primary works, the local authority is constructing a number of adjoining water supply and sewerage installations.

A sewage treatment plant has recently been completed in Skansberget, which is centrally located. This was independent of the installations referred to above.

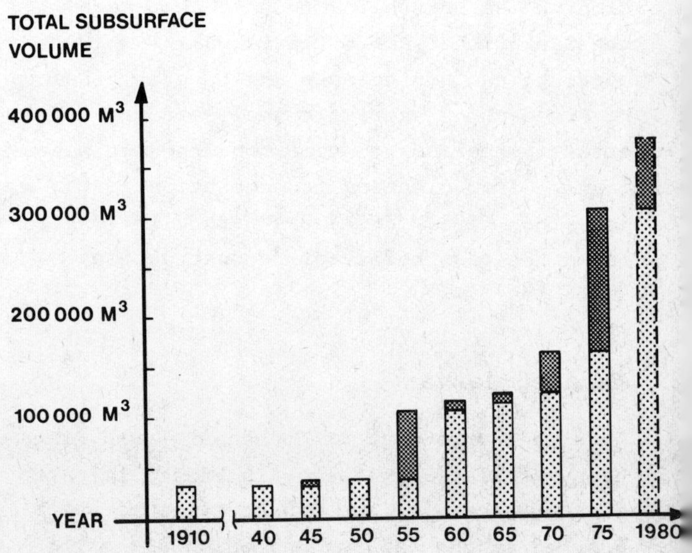

Figure 11
Development of tunnelling in Uddevalla

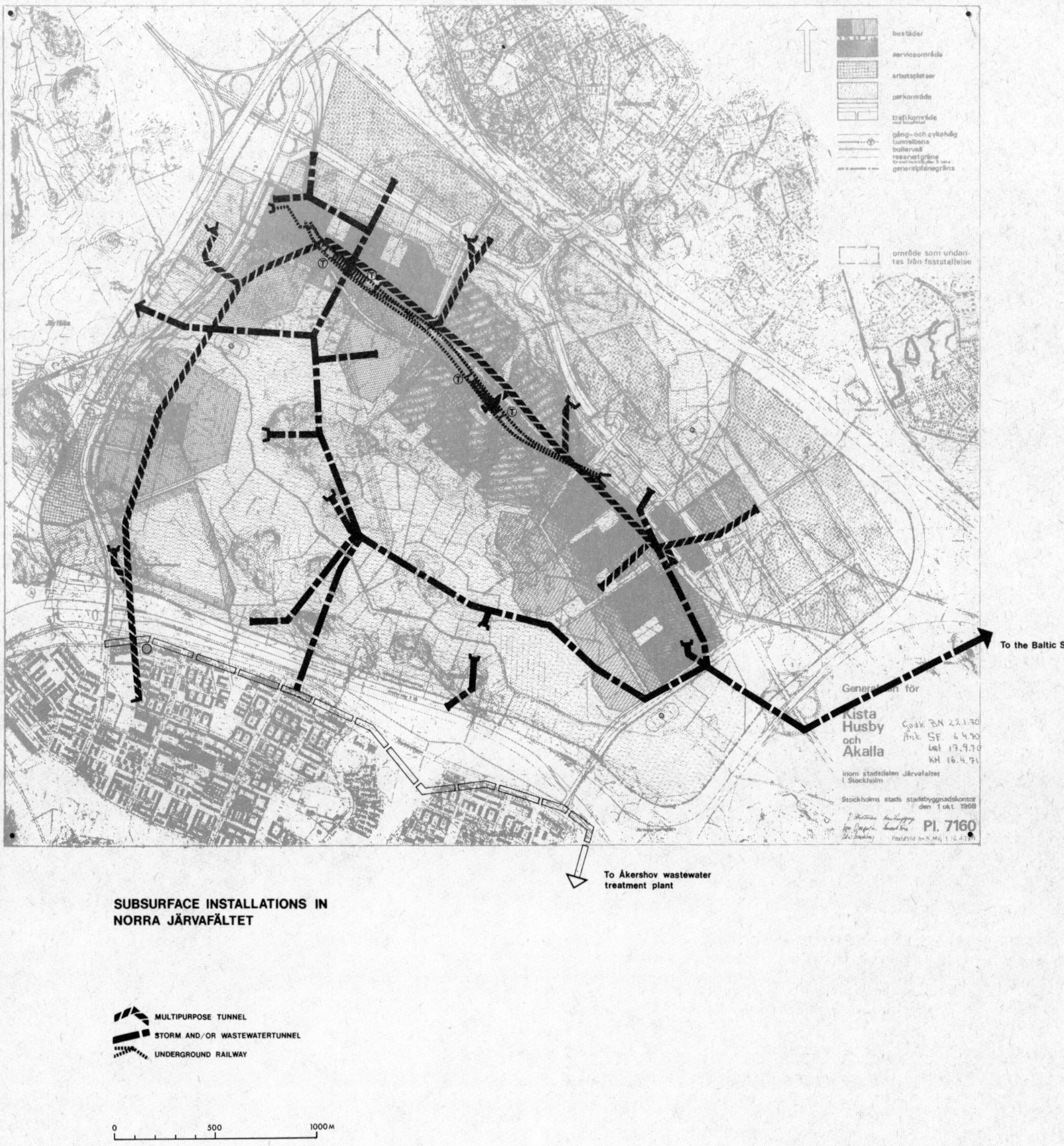

SUBSURFACE INSTALLATIONS IN NORRA JÄRVAFÄLTET

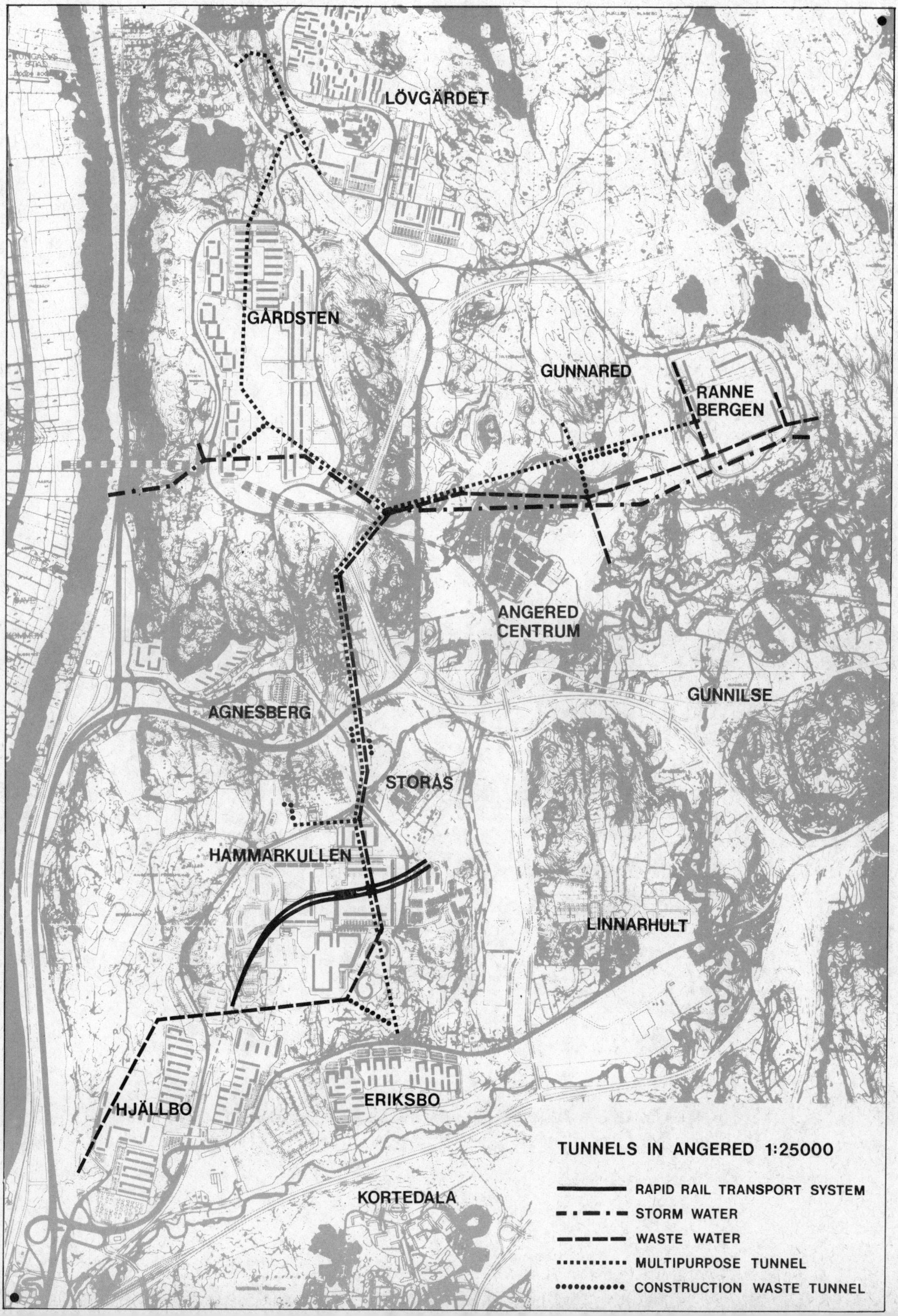
LÖVGÄRDET
GÅRDSTEN
GUNNARED
RANNE
BERGEN
ANGERED
CENTRUM
GUNNILSE
AGNESBERG
STORÅS
HAMMARKULLEN
LINNARHULT
HJÄLLBO
ERIKSBO
KORTEDALA
TUNNELS IN ANGERED 1:25000
RAPID RAIL TRANSPORT SYSTEM
STORM WATER
WASTE WATER
MULTIPURPOSE TUNNEL
CONSTRUCTION WASTE TUNNEL

Number, on map	Installation
1-6	Old railway tunnels
7	Hydroelectric power plant station
8	Military
9	Converter substation
10	Oil storage - grease factory
11	Civil defence installation
12	Sand silo
13	Fresh water tunnel
14-16	Tunnels for water supply and sewerage
17	Bomb shelter
18-19	Railway tunnel
20	Raw water tunnel
21	Waste water treatment plant
22	Tunnel for walkway, cycleway and pipes
23-24	Tunnels for water supply and sewerage
25	Tunnel for diversion of stream
26-27	Tunnels for water supply and sewerage
28	Railway tunnel
29-30	Double motorway tunnel
Installations, under consideration or planned	
31	Discharge tunnel
32	District heating plant
33-34	Bomb shelter
35-36	Oil storage
37	Alternative road tunnel
38	Railway tunnel for industry
39	Parking facilities under housing area
40	Canal locks

Figure 12

A sewage treatment plant located in rock near the centre of Uddevalla. The picture shows those details of the installation which are visible from the outside.

Svenska Vägaktiebolaget

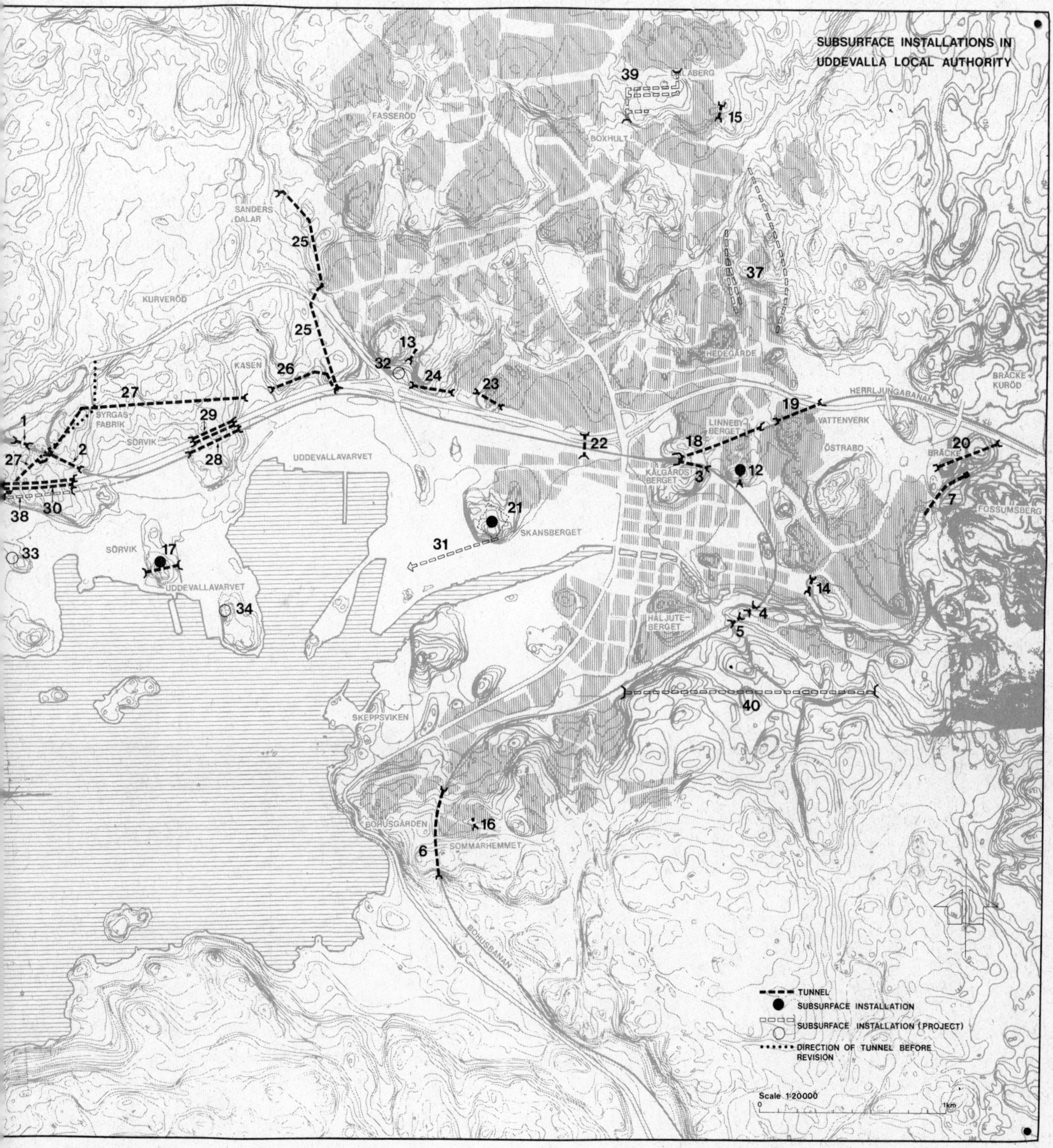
SUBSURFACE INSTALLATIONS IN UDDEVALLA LOCAL AUTHORITY
FASSERÖD
BOXHULT
SANDERS DALAR
KURVERÖD
KASEN
SYRGAS-FABRIK
SÖRVIK
UDDEVALLAVARVET
SKANSBERGET
HEDEGÅRDE
LINNEBY-BERGET
KALGÅRDS-BERGET
VATTENVERK
ÖSTRABO
HERRLJUNGABANAN
BRÄCKE KURÖD
BRÄCKE
FOSSUMSBERG
HÄLJUTE-BERGET
SKEPPSVIKEN
BOHUSGÅRDEN
SOMMARHEMMET
BOHUSBANAN
TUNNEL
SUBSURFACE INSTALLATION
SUBSURFACE INSTALLATION (PROJECT)
DIRECTION OF TUNNEL BEFORE REVISION
Scale 1:20000
0
1km

7. WHY DOES SUBSURFACE CONSTRUCTION REQUIRE PLANNING?

Every new supplement or modification is today subjected to examination in accordance with a number of basic evaluations. These consist of financial, environmental and social consequences. Subsequently, demands are also developing today concerning consideration from energy angle. The responsibility for this consideration rests mainly on the various agencies of the community. The evaluations, however, are of a general character and many companies with responsibility for the development of activitites and installations adapt their approval to suit these higher level demands.

We have tried to answer the question in the heading with four points:

7.1 Utilisation of the subsurface is a valuable resource for urban construction.

7.2 There can be elements of risk combined with subsurface construction.

7.3 The extent of subsurface construction motivates involvement from the community.

7.4 The investigations of the working group show that the production process of subsurface construction of today leaves much to be desired.

7.1 Subsurface installations mean functional advantages

A large number of functions have been considered and implemented in subsurface installations whereby important functional advantages have been noted.

Three important relationships emerge as decisive for the choice of subsurface installations.

Functional needs - the conditions for implementation at ground level

Functional demands - the characteristics of the surrounding subsurface medium

Functional design - the attitude of the human being

Need for supplementary space below the surface of the ground

Urban areas are gradually becoming more densely built-up, along with gradual exploitation of land in these areas for other purposes. At the same time there is an increasing demand for new additions or modifications in order to raise the level of efficiency in the services of and supply systems for the community. The existing utilisation of land, the value of land, and the risks of discomfort and disturbance in connection with new facilities, greatly limit the possibility of meeting urgent demands at ground level.

Topographical obstacles and demands made concerning the conservation of natural and man-made environments make up other limitations on intensified or modified utilisation at ground level.

Interesting examples of subsurface installations which have resulted primarily for this reason are metros with stations (all located subsurface), street tunnels, service line tunnels, transformer and telephone stations and different types of garage and store.

The characteristics of the earth and rock favour the functional demands

Rock surrounding tunnels and rock cavities provides good stability and great durability against different types of influence. Rock provides considerable sealing against leakage, both inwards and outwards. Rock cavities provide exceptionally good protection from observation from outside, external climatic disturbances and damage, including war damage. Rock and earth provide constant temperature and humidity and have a very large capacity to store heat.

At a greater depth below the surface of the earth, the rock is saturated with water under pressure which corresponds to the height of the ground-water level located above it. The water and the water pressure prevent the leakage from rock cavities of gases or liquids stored there.

Almost independent industrial construction activity has been developed in Sweden for the storage of oil and other petroleum products in rock cavities based on this characteristic.

Other important applications for the capacities of rock include air raid shelters, defence and civil defence functions, war-protected industrial activities, refrigerator and freezer installations, water and sewage treatment processes, different types of energy storage and also the storage of sensitive, hazardous and environmentally dangerous substances and products.

The conditions for working and spending time below ground are favourable

The possibility of extending subsurface installations independent of the use of and limitations at ground level provides possibilities for individuals to move fast in transport systems and to reach destinations free from interruption and disturbance at ground level.

Subsurface installations provide a constant climate for people who work and spend their time there. They are also completely free from interruption and disturbance due to traffic and activities at ground level.

Metros with stations and service facilities as well as pedestrian tunnels have been created and accepted as facilities which favour the movement of city inhabitants in inner-city areas.

Tunnels for different types of service system provide a good working environment for installation, maintenance, repair and control work. The working environment in treatment plants located subsurface is much more favourable than in corresponding facilities at ground level.

The subsurface is a resource that must be exploited wisely

By way of summary it is important to state

- that the subsurface constitutes a resource that must be considered when there is lack of space and/or other limitations at ground level,
- that subsurface cavities in rock and soil have characteristics which favour various functions and
- that subsurface installations offer favourable conditions for people to work and spend their time in. Subsurface location should be considered both in planning of particular functions and in overall planning in order that the subsurface is exploited in the best possible way.

Figure 13

Interior of multi-use tunnel in Angered, Gothenburg. On the left, cables for electricity and telecommunications, in the middle fresh water pipes and to the right foundations for district heating pipes.

Skanska

7.2 Subsurface installations can cause disturbance and damage

During the construction stage

The construction of tunnels and rock cavities implies noise and vibration from blasting. The removal of the excavated stone requires transport which affects the environment. Regulations concerning supervision by the community during the construction stage, however, make it possible to keep this disturbance within acceptable limits. Information on ongoing subsurface work provided for people in the vicinity is generally insufficient, which causes anxiety and inconvenience.

Rockfalls etc. affecting the environment are rare. Cases have occurred, however, where information on existing rock cavities is so bad that tunnelling work has broken into existing rock cavities.

The leakage of water into tunnels during the construction stage is a normal occurrence. After sealing work, this does not need to have any effect on the environment. In certain cases, however, reservoirs of ground water are perforated and this implies permanent settling in the layers above the rock cavity even if leakage into the tunnel or rock cavity ceases. There are no general requirements on control of leakage into tunnels and cavities or on analysis of the consequences of such leakage.

During operation

The dominant risk of disturbance and damage to the environment during the construction stage is also associated with water leaking into the tunnel or cavity. Even very moderate leakage can result in a lowering of the ground-water level in a relatively extensive area around the subsurface installation.

Rock does not seal completely against the leaking out of gas. If adequate sealing has not been applied, petroleum gases, for example, from an oil storage installation can seep out and even reach the surface of the ground.

The stability of the rock round a cavity requires an equal amount of attention from the subsurface installation itself and from construction work being carried out in the vicinity and having its foundations in the immediate vicinity.

Similar attention is required in the case of extreme temperatures in the rock cavity.

Installations can have been created below ground level without critical consideration of the environment in which people work or spend their time, so that people are obliged to work and spend their time in the subsurface under unnecessarily unfavourable conditions.

A subsurface installation may be valuable as a resource for the community but a badly built subsurface installation is just as dangerous to the community it is designed to serve. A conduit tunnel in rock has been chosen in order to preserve a highly valued environment in a city. Through inefficient grouting of the rock or failure to take other measures, ground water seeps into the tunnel so that the ground-water level is drawn down. Foundations and buildings in the environment to be preserved are endangered and can suffer irreparable damage.

Disturbance and damage to land and buildings from a subsurface installation can cause considerable supplementary costs. It is the owner of the installation who suffers primarily. It is by no means infrequent that it is the community that is under obligation to compensate individuals who have suffered.

Figure 14 - 15

Public air-raid shelter in central Stockholm.

In peacetime, the rock cavity is used for parking and servicing facilities for cars.

7.3 The responsibility of the community

It is the responsibility of the community to ensure that disturbance and damage do not affect individuals or the general public in connection with subsurface installations. An important factor in this surveillance is close supervision of planning and analysis work so that available knowledge is obtained of the preconditions for the subsurface installation and so that the necessary techniques are used.

The community assumes in other similar contexts responsibility for ensuring that installations are efficiently utilised. Co-ordination for multi-use of tunnels is an example of this obligation.

It is considered that the co-ordinated evaluation of economic, environmental and social points of view attesting to the advantage of a subsurface installation showed also demand the involvement of the community.

In summary it is considered that subsurface construction and its consequences satisfy most of the criteria laid down in the Swedish Building Law and the Environmental Protection Law as motives for participation by the community in planning, construction and control work.

7.4 What are the defects today?

Material forming the basis for the position of the working group

It has been possible to base an appraisal of the advantages and disadvantages characteristic of current procedures for planning and administrative handling on several types of material.

- The interviews carried out in eight specially investigated communities.
- Detailed analysis of the three areas Angered, Kista-Husby-Akalla and Uddevalla. This analysis provides in different ways considerably clearer illumination of the procedures than the interview surveys. Many projects have been subjected to detailed study. There is documentation of several measures taken, the number of interviews and personal contacts is very large.
- A large number of people active in and with experience of subsurface construction and planning have been contacted and this has provided even more basic material for observations.

7.4.1 General views

One general impression obtained from the interviews carried out in the smaller local authority administrations is that problems involving subsurface installations and the planning of these do not generally give rise to deep involvement. Nor, in many cases, do they spontaneously inspire requests about or demands on measures to be taken. Different local authorities have, in different ways, found models for carrying out their planning and co-ordination in an acceptable way.

Opinions expressed in Stockholm and Gothenburg indicate that attempts have been made to meet the demands of subsurface construction within the local authority administrations. The considerably greater extent and complexity in subsurface construction in the large cities has, however, resulted in experience which considerably underlines the need for better and more goal-oriented planning for the utilisation of the subsurface.

National and local authorities are also responsible for the construction of a good deal of the most extensive subsurface installations. Conflicts of interest can easily occur in the management and decisions of the authorities concerned. Rapid and low-cost implementation stands in opposition to careful planning with open information and debate.

Figure 16

Compared with surface tankage the storage in rock caverns of petroleum products offers advantages: lower construction costs (when the volume exceeds 25-40 000 m³), superior protection against fire and significantly lower operation costs.

Svenska Vägaktiebolaget

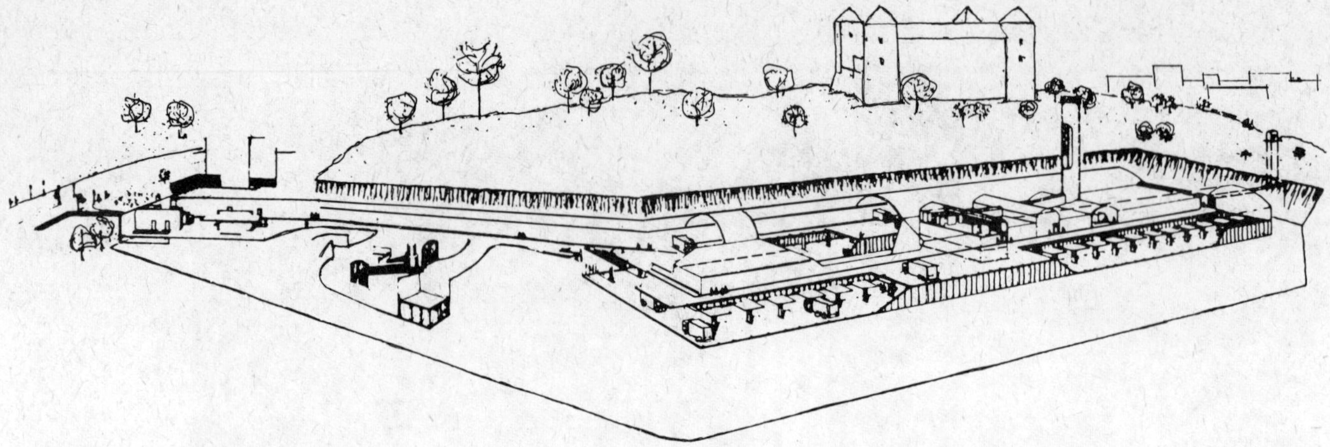

A public cleaning firm has solved its garage problems by constructing a parking space of 100 000 m³ in the rock under central Stockholm. The alternative would have been to locate the garage at a considerable distance from the centre. The first part consisted of constructing a stone crusher in the rock in which the rocks excavated later on could be crushed.

Sellbergs

According to the impression gained by the people carrying out the survey, it is obvious that the full range implied by the problems involved in subsurface construction and planning is relatively unknown among those active in local authority administrations and, to a certain extent, also among national authorities and those active in subsurface construction.

7.4.2 Uncertainty concerning the responsibility of the local authorities

The prevailing degree of uncertainty among all local authorities and people interviewed concerns responsibility for the planned utilisation of the subsurface and for the administrative handling of planning and construction of subsurface installations.

At present, no local authority has general obligation or right to plan or to take responsibility for the creation of subsurface rock cavities.

a) Planning of particular functions is not carried out in order to indicate subsurface utilisation on either a short- or long-term basis. Overall planning, mainly the responsibility of the local authority does not generally for this reason take into account subsurface construction and the consequences of this.

b) In certain communities subsurface installations are described in detail plans but in many they are not described at all. The descriptions vary to a great extent. Consequence analysis is generally conspicuous by its absence. Installations located on land owned by the local authority are often not described at all. Security requirements eliminate every form of description in certain cases.

c) Notification is made in certain cases to the Building Committee or applications are made for building permits, without, however, a check being carried out of the situation and shape of the rock cavity or of the consequences to the environment. Consideration of the suitability of a rock cavity is often outside the competence of the Building Committee.

d) Stockholm and Gothenburg, and partly also Lidingö and Lysekil, have better developed routines in the handling of subsurface matters. For example, individuals and groups have been allocated responsibility for the registration and planning of subsurface installations. Co-ordination tasks have often been allocated to street maintenance and building plan-offices, sometimes to the Local Building Authority. The geo-information which is so very important to planning and construction work is not compiled. Often, all information of this kind is spread out in the hands of whoever has commissioned the work, authorities and consultants.

c) In spite of the lack of special suggestions or support in instructions, certain communities have found it important to carry out joint planning of activities above and below ground level. Other communities do not even consider this as a need at all.

7.4.3 Defects - conflicts - damage

Planning

As a direct consequence of uncertainty concerning local authority responsibility with respect to subsurface construction and planning, the following defects can be noted in the utilisation of the subsurface. These defects can cause unnecessary costs and unnecessary spread of inconvenience.

Subsurface installations influence construction work above surface and, conversely, construction above surface influences subsurface construction. However, no co-ordination has been carried out between those responsible for the respective activities.

Planning above surface is carried out without those responsible being aware of the fact that subsurface requirements exist or can be anticipated. Planning of particular functions which should indicate subsurface utilisation does not keep pace with planning of construction above surface and no consideration is being given to subsurface siting and subsurface projects in such good time that co-ordination and mutual consideration can be applied between construction work above and below ground level.

a) Subsurface installation requires a safety zone and a protective zone in order to avoid endangering stability. This is a fact wellknown to technologists but almost completely unknown to those responsible for planning.

b) Even relatively moderate leakage of ground water can have a great affect on the foundations of buildings above ground level. Conversely, a constant ground water table can affect the possibility of building subsurface facilities.

Multi-use installations

Different subsurface needs result in separate subsurface installations although multi-use should have been considered.

a) Multi-use has not been considered due to the subdivision of responsibility for different functions. No co-ordination has been carried out.

b) Multi-use has been discussed but without any considered goals or detailed studies.

c) Multi-use has been rejected due to incompatible functions.

Crowding and conflicts

In an area with favourable conditions but limited in extent, various subsurface needs require space, either on one single occasion or over a more extensive period of time. As a result, crowding or direct collisions of interest can occur for the following reasons:

a) The need in urban development for subsurface accommodation has not been considered from the points of view of planning, co-ordination or the long period.

b) Subsurface projects have developed as a result of typical short-term thinking.

7.4.4 Implementation problems

Partly due to inadequate local authority responsibility, partly due to the lack of general regulations on planning, preliminary studies and design work on subsurface installations, the following defects can be noted.

Registration

The position and shape of existing subsurface installations is not sufficiently well-

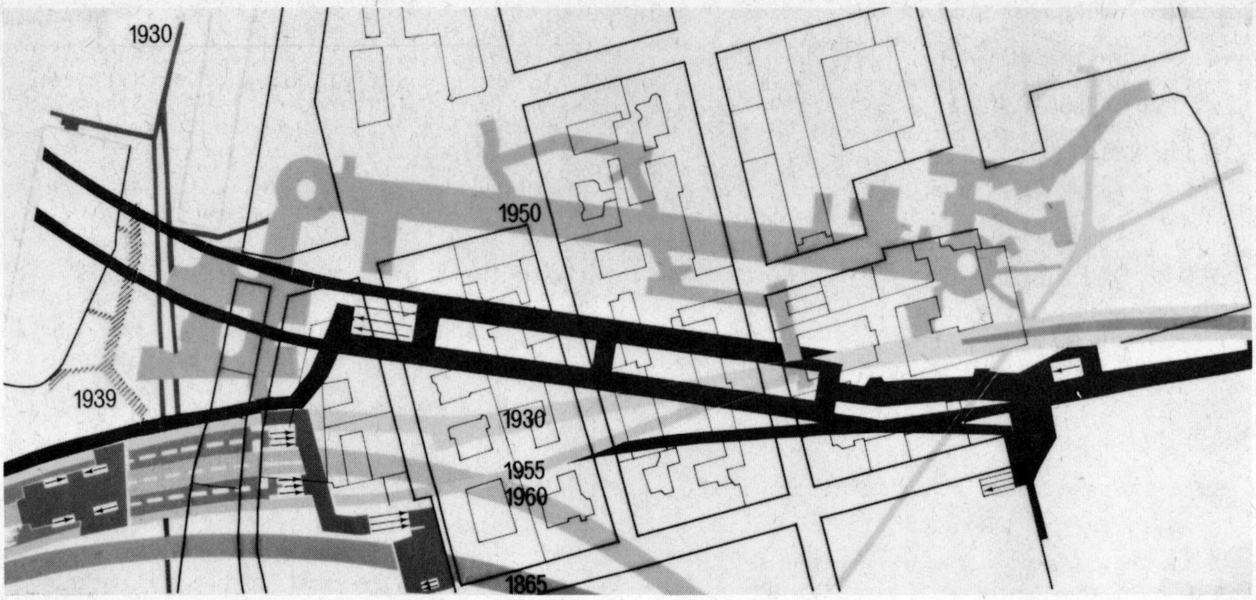

Figure 17

An example of an area in central Stockholm where shortage of subsurface space has resulted in complicated subsurface installations.

known and this can lead to direct conflicts in construction work. The reasons for this are:

a) The community does not register subsurface installations systematically and lacks information on older (and more recent) installations.
b) Information in the community is difficult to find and is not used when examining projects or in the processes of decision-making.
c) The information is limited for security reasons.

Geo-information

The results of old geo-investigations in the area where subsurface installations are to be located is not known or are difficult to find, as a result of which valuable and expensive surveys cannot be re-used.

a) No information is available within the community
b) Information within the community is both dispersed and unprocessed.
c) The information is subject to security regulations.

Completed subsurface installations have seldom been plotted from the standpoint of rock geology. There is often no information about reinforcement and observations with respect to water leaking in. This means that a qualitatively superior source of information cannot be utilised from the time when the completed installation was accessible and available for observation.

Preliminary studies - consequence analysis

The consequences of damage to the environment in the form of noise, shaking, lowering of the ground-water level, settling, landslides, disturbance caused by transport and large stocks of materials are not sufficiently known.

The reason for this is that:

a) No studies have been carried out where the risk aspect is clarified.
b) Studies have been carried out but the risks have not been calculated.

c) The risks have been calculated and consciously taken into account but have not been passed on to decision-makers and those living in the vicinity.

The technique of carrying out surveys to determine rock quality and ground-water conditions beforehand is well-developed. Information on these possibilities appears, however, not to have been well-distributed. Mistaken thrift often results in the carrying out of all too feeble surveys since all too great a consideration is given to the risk of incurring extra costs in construction work. It is, however, natural for subsurface construction in rock to imply a certain amount of risk of unexpected problems during the construction stage.

Rockfalls, damage, formblasting and the risk of inundation through water leakage occur in spite of the fact that well-developed survey techniques are available. Cost calculations are wrong due to unexpected reinforcement requirements which could have been anticipated with a considerable amount of certainty.

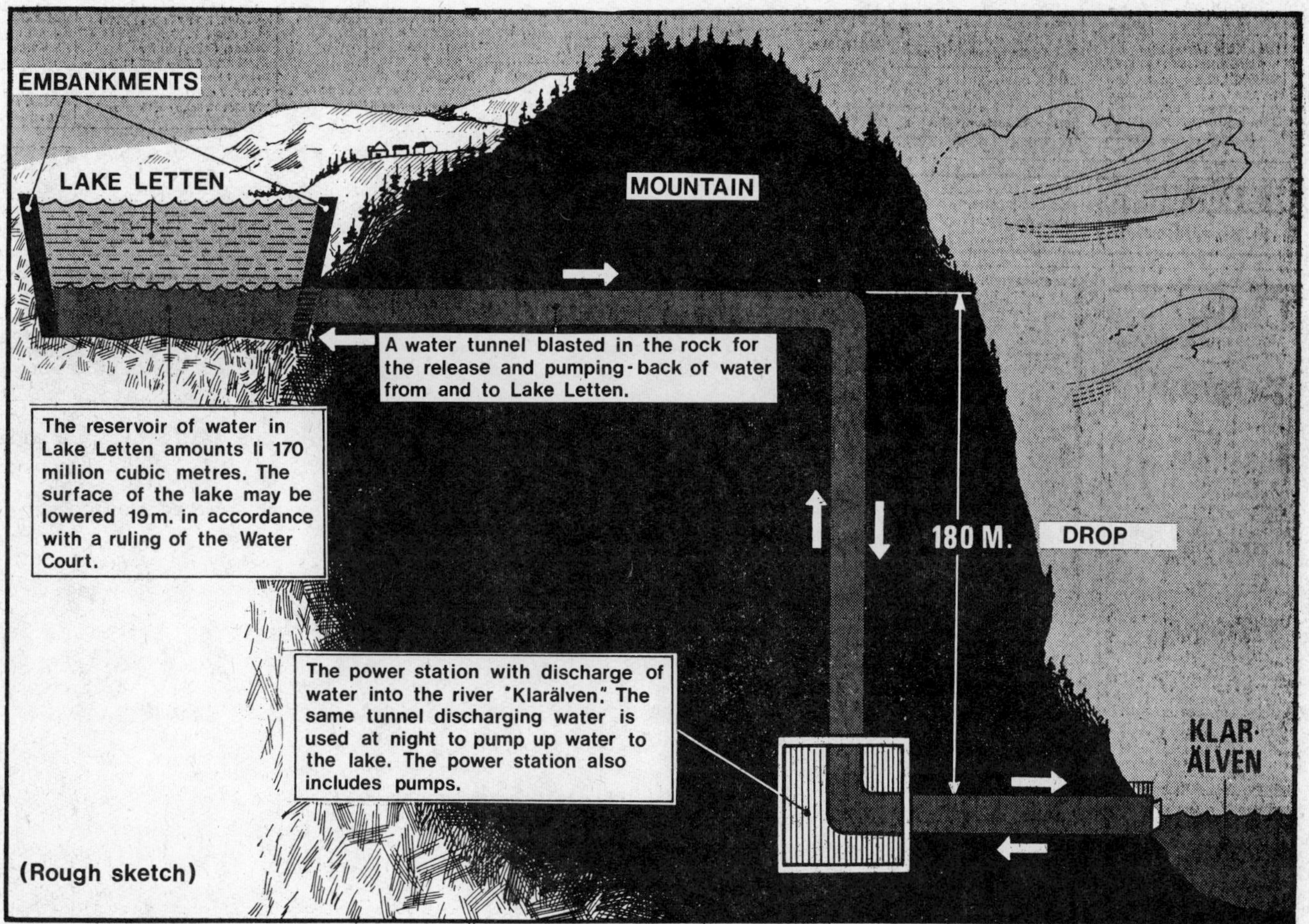

Teckning: GÖTE BERGMAN

Figure 18

Pumped storage power stations, the purpose of which is to even out the consumption of energy, may be expected to be installed to an increasingly greater extent. One common method is, as shown here, to connect by tunnel two reservoirs situated at different levels. There is a proposal however that the lower reservoir be excavated deep under the surface. This solution offers great freedom of location.

a) The damage which occurred could have been reduced or avoided if a comprehensive analysis of ground and rock conditions had been carried out before preliminary design work was started.

b) Classification of the risk of damage would have occurred if detailed surveys with suitable drilling work had been carried out.

Cost-benefit analysis

Evaluation of the relationship between benefit and cost has to a very limited extent made up the basis for decision on subsurface installations. The reasons for this are deficient awareness of the problems in subsurface construction work, but also of the possibilities of solving functional problems below ground level.

The lack of comprehensive cost-benefit analyses can therefore have resulted in:

a) Subsurface location being chosen where a position above ground level was preferable.

b) Position above ground level being chosen where subsurface location would have been more advantageous.

The lack of information about subsurface construction in general and about specific installations, as well as defective planning, where time after time unnecessary damage or disturbance has been caused are tend to create suspicion concerning subsurface construction technology and the possible utilisation of the subsurface.

7.4.5 The subsurface as an environment for human beings

The characteristics of the subsurface which are decisive for the way in which they are experienced consist of lack of daylight, limited evaluation possibilities, and the climate, which is often completely artificial. The subsurface has these characteristics in common with other premises such as the interior and basement parts of buildings above ground level. Defects in supervision of subsurface design and the lack of general analysis of the social integration and dependability of the subsurface tends to give it a low social status. Scientific results hitherto published concerning the physiological and psychological reactions of human beings give only partial support this under-evaluation.

This last-mentioned observation, however, does not imply that accommodation above ground level with access to daylight, ground surface contact, etc., must be accorded a higher quality value as an environment for human beings.

8. WHAT PLANNING SHOULD BE CARRIED OUT?

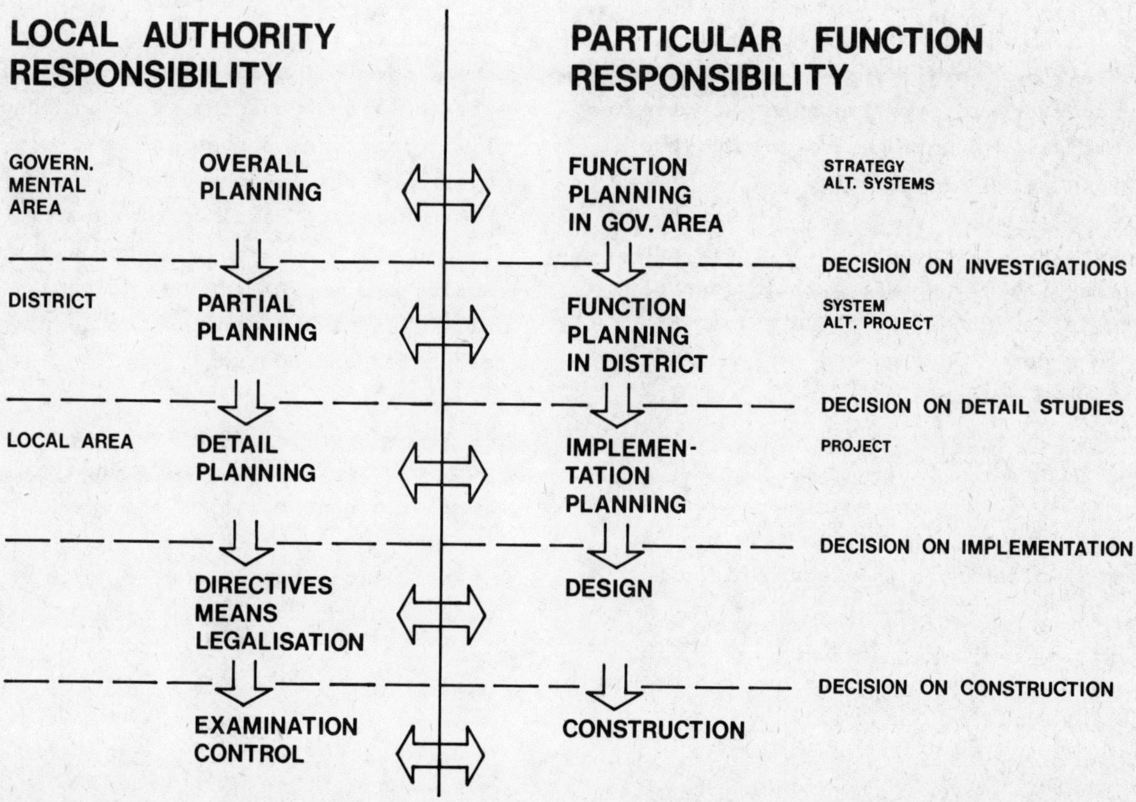

In the previous section an account was given as to why subsurface construction needs to be integrated with community planning. How can this be realised?

Three types of planning have earlier been described:

Detail planning - aimed at planning surveys, studies and measures to implement individual and limited projects.

Planning of particular functions - aimed at planning, on a short- and long-term basis, for the needs and demands of a subsidiary function with respect to the associated systems and individual projects.

Overall planning - intended to provide guide-lines for community development as a whole, to co-ordinate different demands from a planning perspective and to supervise the rational utilisation of resources.

Detail planning is central to the expert implementation of a subsurface installation. Detailed knowledge concerning conditions for planning the installation is essential if the necessary integration into the planning of particular functions and in the overall planning is to be possible.

In a fully comprehensive planning system, as described below, detail planning can be found as a part of the third stage which precedes detail design and construction. The system consists of three principle stages:

- overall planning - long-term planning
- partial planning - medium-term planning
- detail planning - short-term planning

Overall planning and planning for particular functions are assumed to be carried out in parallel and in co-ordination.

The following table describes the objectives of co-ordinated planning at the three stages:

	overall planning	partial planning	detail planning
subsurface solution	alternate systems for reservation of resources	alternate studies for decision on location and solution	pre-design for decision on final design and construction
topography geotechnique	general review and complementory testing	pre-investigation for evaluation of techniques	detailed investigations for design and construction
cost-benefit	general calculation	preliminary cost-benefit analysis, preliminary study of consequences	final cost-benefit analysis, description of consequences

Detail planning and the technique developed from it of constructing subsurface installations have, through research and development work, reached a potentially very advanced level in Sweden. When this well-developed knowledge is not always made use of, particularly in urban construction work, this can be explained to a great extent by the fact that those commissioning the work and those making the decisions consciously or unconsciously limit important survey and analysis work. No system of regulations is available for the administrative handling of subsurface installation which have been sanctioned by the community and which give the community the possibility of making demands on planning, construction and control.

In the following, this report will devote interest primarily to the demands made by subsurface construction on the community and the way in which the planning and design of installations could be co-ordinated with the overall planning of the community.

8.1 Detail planning

8.1.1 The present routine

The implementaion of numerous subsurface projects has gradually produced routines which can be described as general. The most important stages in project development are the following.

Planning

- Analysis concerning the demands made on the installation.

- Preliminary survey of the geo-conditions for the choice of location.

- Preliminary design work on the installation, including appraisal of environmental influence.

- Possible consideration of permits if there are any legal considerations or concessionary obligations involved.

Design work

- Detailed survey of the geo-conditions in the fixed situation or for the determination of position.

- Detailed design of the installation including general reinforcement and sealing work.

- Inspection of adjacent buildings and wells, possible programmes for measuring ground water.

During construction work

- In situ surveys of geo-conditions and geoplotting.

- Detailed instructions concerning reinforcement and sealing.

- Plotting of positions, reinforcements and sealing.

It is difficult to draw any fixed boundary line between planning and design work. One criterion for planning consists of the measures which precede the decision concerning permits and the decision made to build the installation.

Construction responsibility normally covers only the construction of the installation. In the case of turnkey negotiations, however, the construction responsibility for an installation can also include design work.

8.1.2 Demands made on planning

The type of subsurface installation makes different demands on planning. A few examples are given below.

Storage of oil for readiness in case of emergency requires that a national strategy for oil storage be made up, that this storage strategy be integrated in community planning and the infrastructure in the field of transport, that geology be analysed with respect to consideration of the demands made on the storage installation and that the effects on the environment be examined. Consideration should be carried out at national, regional and local authority levels.

A regional sewage treatment plant subsurface requires deliberation at regional and local authority levels. Consideration has been given to existing and future buildings and also to the possibility of connecting the plant to existing and future sewer systems. The environmental effects of various alternative sitings have been examined. The geological conditions have been analysed with respect to both technique and cost.

An automatic telephone exchange at subsurface level generally requires a position in a relatively central area of the city. Apart from analysis of the geological conditions, it is necessary for studies to have been carried out of existing and future buildings and also of their foundations, in order to decide on a suitable location and to ensure protection of the environment during the construction and operating stages.

A road tunnel usually has its location determined by the geometry of traffic arteries. Subsurface location can not, in other words, be regulated by the geological conditions. In planning for this, it is necessary for the geological problems which occur to be thoroughly investigated and measures carefully considered to prevent interruptions in the tunnelling work and disturbance to the surrounding ground and buildings.

This should indicate that subsurface location of various types of installation should be given attention at both national and regional levels as well as at local authority level, if it is to be possible to acquire complete confidence that the installation has been planned with all the careful consideration required by the community. It is easy to come to the associated conclusion that demands, reservations of area and plants should be covered by the system of regulations within the Building Law, independent of whether demand is concerned with location, above surface or subsurface.

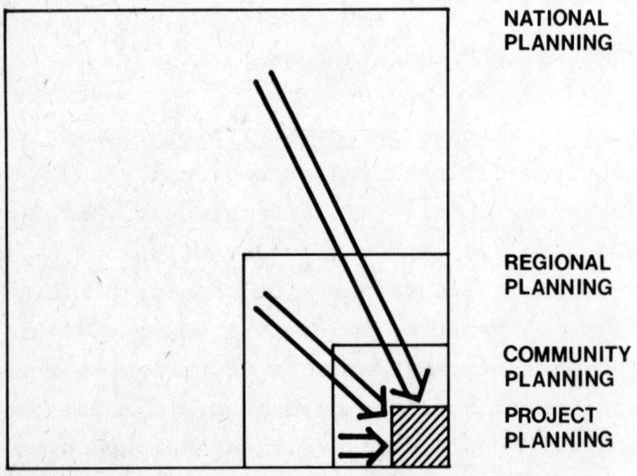

It is basic in detail planning that it is ensured in advance that the utilisation of the completed installation corresponds to the cost involved. This means that all reasonable necessary information must be collected and that the highest level of expertise and experience must be used to appraise the consequences during the construction stage and during utilisation of the completed installation. Particular attention is to be aimed at appraisal of the risk of unexpected incidents and defects.

Experience has taught us that the possibilities of influencing a project are much greater at an early stage than at a late stage. The cost of efforts which can have an effect on the project are also considerably lower at an early stage.

Experience of subsurface construction indicates that temporary cessation of work during tunnelling or during the construction of rock cavities can have a serious effect from the point of view of increasing costs.

Preliminary surveys of geo-conditions should therefore be aimed at plotting the risk of unexpected structural deviations. The parties involved in the planning and implementation of subsurface installations must, however, always be conscious of the fact that, for financial reasons, surveys can seldom be exhaustive. The responsibility for unexpected difficulties can often, therefore, not be laid on those responsible for surveys, planning and construction. Confident cooperation for the solving of problems that occur is particularly necessary in connection with subsurface construction.

Disturbance and damage to land and buildings in the environment often result in demands involving high compensation.

For this reason great efforts should be devoted to the survey and appraisal of geology and hydro-geology in order to avoid weak zones, leakage, etc. or to counteract the effects of such factors.

It is relatively simple to specify the utility value of a subsurface installation from the point of view of direct financial advantage for the person owning it. In spite of this, there is a shortage of collected in-

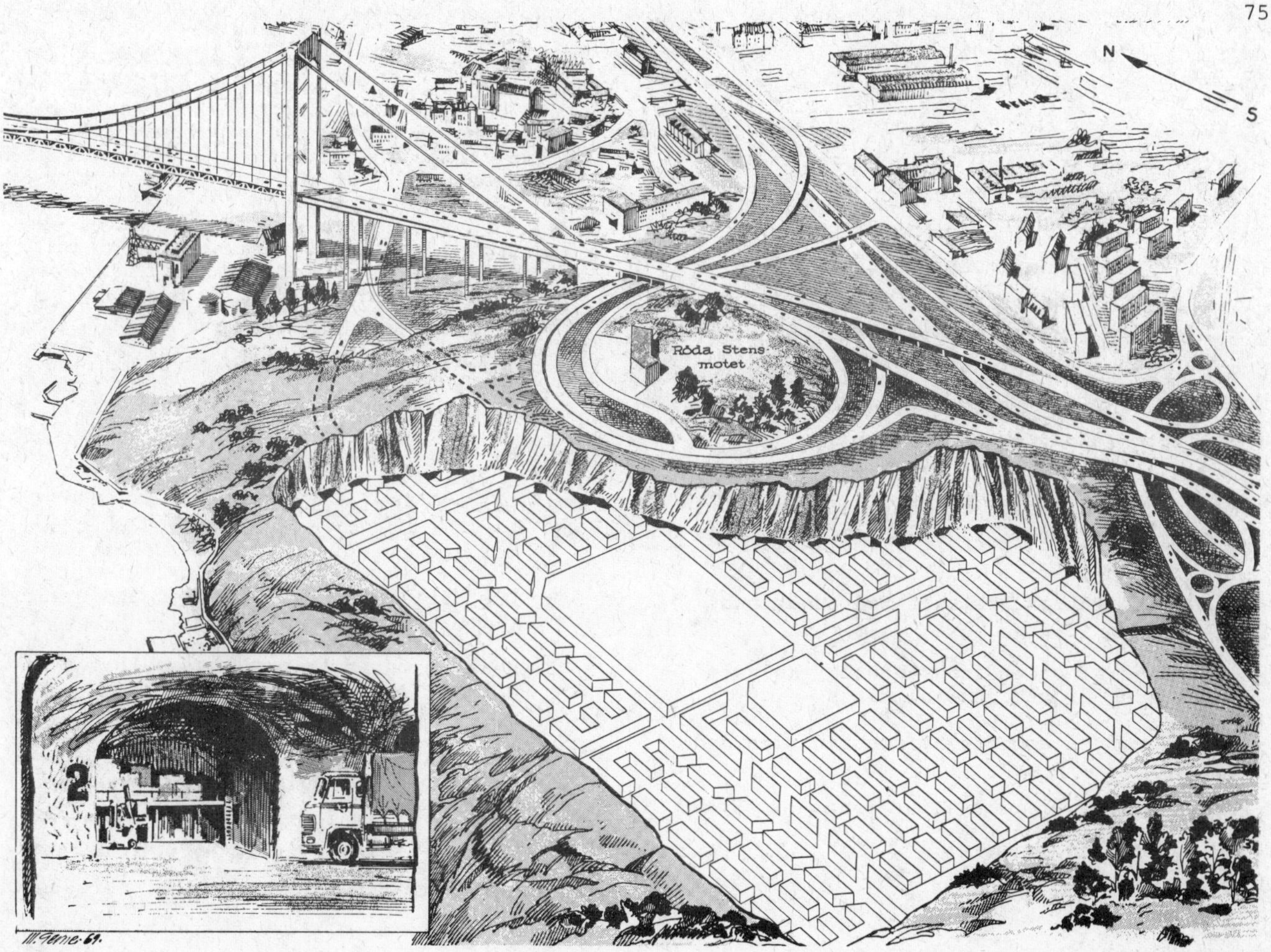

Figure 19

Proposal for a large subsurface installation beneath central Gothenburg, with the aim of housing different kinds of industrial activity, storage etc. The shape of the different subsurface areas can be planned in detail in accordance with the wishes of the user. Subsurface location makes possible the utilisation of a unique site which has direct connections with the port, the railway and the European road network.

formation which would make possible more general evaluation for most types of installation. Utility from the point of view of an environment where people work and spend their time, and also the effects on the environment have hitherto not been subjected to systematic processing. This deficiency depends to a great extent on the meagre interest shown in subsurface construction by the community and those using the installations.

As a general observation, it is important for the detail planning carried out to be well-developed, particularly with respect to the demands made by different functions, but the preparatory analysis has certain defects, partly connected with the extent and accuracy of the preliminary survey, and partly in co-ordination with overall community planning.

In order to improve detail planning so that it corresponde to the demands otherwise made by the community on activities and construction:

1. planning and construction of subsurface installations should be integrated into overall community planning.

2. requirements should be set up, concerning the surveys and studies particularly needed by subsurface construction.

3. administrative routines should be developed in order to meet the demands made by subsurface construction.

If we assume that the planning of subsurface installations is to follow the systems of regulations and routines of the Swedish Building Law, this means that a project which is subjected to detail planning must be covered by the town plan or must be subjected to corresponding examination. The project must also be described in the information routine which applies in the case of the town plan for land-owners, other parties in the case, authorities and the general public.

Detail planning must therefore aim at providing the following basic information:

- a requirement analysis of the functional contents
- geo-information and consequence analysis concerning ground and buildings
- environmental effects of construction and operation
- demands made on protective areas
- cost-benefit appraisal

8.2 Planning of particular functions

In the planning of particular functions, which primarily applies to technical-economical considerations concerning particular functions, there is often good reason to initiate subsurface location of installations. Here therefore, particular attention should be paid to the conditions for subsurface location.

In general, when planning particular functions one should

a) have access to the necessary relevant information about techniques, costs and characteristics of the subsurface

b) make an unbiased assessment of the functional requirements in subsurface location

c) acquire the best possible information on the geo-conditions and the environment which could be affected by subsurface installations

d) make possible co-ordination of the planning of particular functions with overall planning, and also the required participation of the general public in the decision process

e) draw up the best possible account of subsurface installations related to known reference points, and carry out the best possible consequence analysis regarding the environment

f) attempt to cover the investigation headings, applicable to different types of planning, such as:

A Formulation of problems and goals

B Listing of preconditions

C Working out of possible alternatives

D Description of consequences and evaluation

E Working out of proposals and plans of action

Particular function planning has so far been aimed at covering relatively short-term needs. It is desirable that assessments are extended to apply even to medium- and long-term requirements so that more hypothetical requirements can be taken into consideration in coordinated overall planning.

8.2.1 Partial planning of particular functions

The problem of subsurface location of particular functions means that partial planning of a mainly overall character from a medium- and short-term point of view, is the most important level of planning. On this level there are conditions leading to the acquirement of more detailed information on geo-conditions and the environment. Furthermore, there are possibilities for making a more quantifiable cost-benefit analysis, which gives a basis for decisions on the implementation of subsurface construction.

When planning on this level one should

a) make an inventory of and summarise the necessary geo-information (preliminary investigation)

b) carry out preliminary design of subsurface alternatives

c) prepare consequence analyses

d) carry out a cost-benefit analysis for evaluation

e) on the basis of these studies describe subsurface installations which are necessary from a <u>medium-term</u> point of view and account for subsurface projects which should be decided on for implementation from a <u>short-term</u> point of view (detail planning)

8.2.2 Overall planning of particular functions

The planning of particular functions covering communities and regions, which aims at clarifying long-, medium- and short-term requirements, must usually be hypothetical to a great extent when taking into consideration the long-term need for subsurface space. In spite of this, assessments are of the greatest significance to enable the preparation of the co-ordination of different types of overall planning.

Planning on this level should:

a) consider and concentrate attention on all possible subsurface requirements from a <u>long-term</u> point of view, concerning both extent and problems

b) describe the demands on time and location of definable subsurface requirements from a <u>medium-term</u> point of view

c) give an account of subsurface projects which have been prepared for implementation from a <u>short-term</u> point of view.

8.3 Overall planning

8.3.1 The need for rules in overall planning for subsurface construction

The overall planning being carried out today is aimed at co-ordinating various types of requirement and awarding priority to or limiting utilisation of resources within communities or regions. This type of planning can be expected to increase in extent.

A description has been included earlier of the sectors which together make up the overall planning of the community. They are

Joint planning conditions
Finance
Social planning

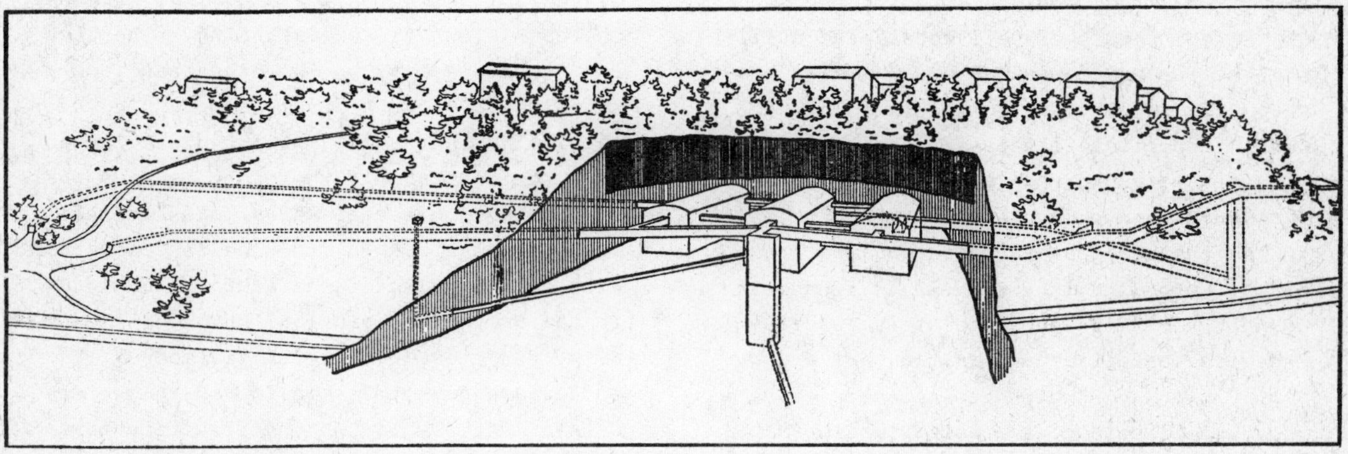

Figure 20

The picture shows one of the major telephone exhanges of the National Telecommunications Service, located in an urban area. Because of its location maximum protection against sabotage and acts of war is achieved.

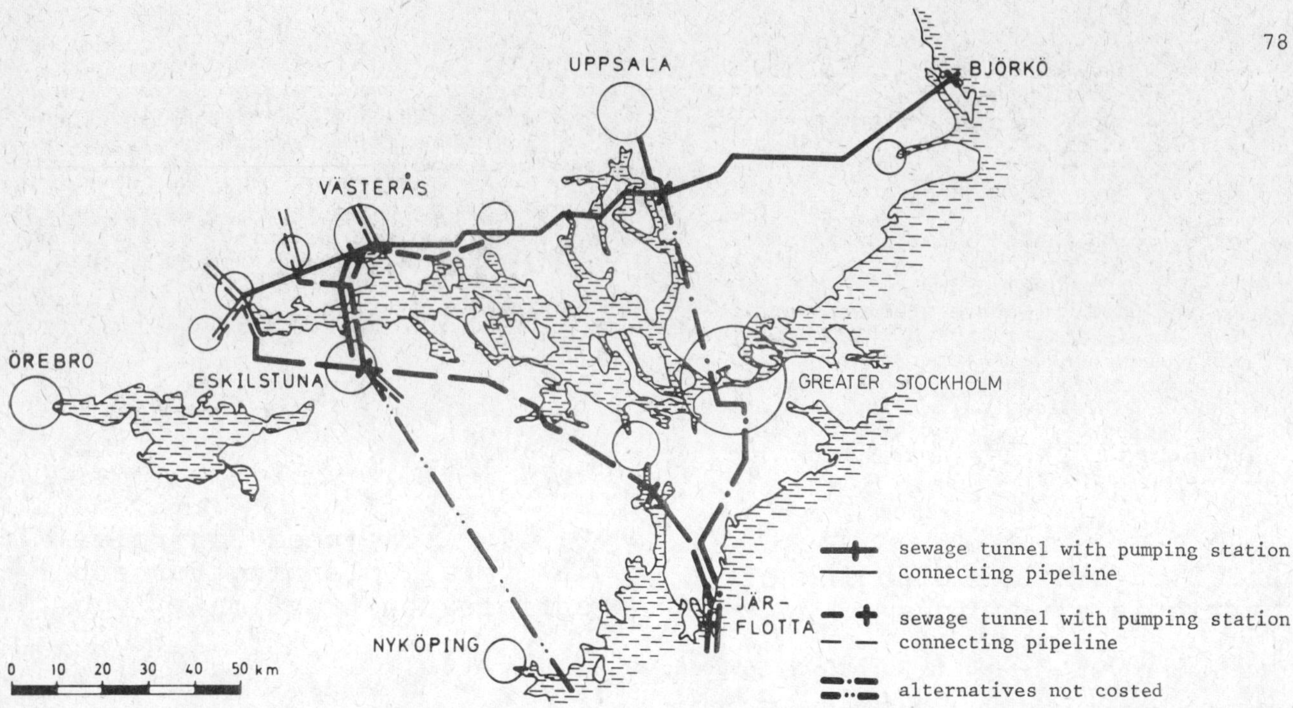

Figure 21

The 1966 preliminary outline for a regional plan for the Stockholm area took up the idea of a regional sewerage system to cover the area from the inner reaches of Lake Mälaren to the Baltic Sea. There would then be practically no discharge of sewage into Lake Mälaren. An investigatory study was carried, the results of which showed that the project will not be realised within the foreseeable future.

Physical planning

Energy

Natural resources

The making up of joint planning conditions constitutes the combined form for the co-ordinated objective, well-considered utilisation of resources and a programme of measures for development in the immediate future. Important sectors for the utilisation of the subsurface and of subsurface construction are physical planning, that is to say the utilisation of ground and functional co-ordination, as well as financial planning. Other sectors closely associated with subsurface utilisation are energy supply planning (which in this case means resource planning), the determination of consumption distribution and standard levels as well as the planning of natural resources. The social planning sector, meaning relationships between the individual, the group, the well-being of the community, and culture, is, paradoxically, not the object of methodical planning. However, problems associated with time spent by people below ground level should be viewed as a planning task within this sector.

Overall planning, on the one hand, provides the conditions for the planning of particular functions but, on the other hand, obtains basic material from the planning of particular functions. During overall planning, demands should therefore be made whereby the basic material which makes overall planning possible is produced in connection with the planning of particular functions. In the overall physical planning of a community, for example, the utilisation of the subsurface cannot be considered unless necessary information on the requirements and requests from planning for particular functions has been given.

Important tasks in overall planning include consideration of questions at various planning levels concerning the multi-use of subsurface cavities and the co-ordination of above surface planning with subsurface planning. One generally interesting example of such co-ordination is consideration of existing and future subsurface installations for civil defence purposes. According to a suggestion recently submitted concerned with guide-lines for civil defence air raid shelters, this obligation may rest with the local authorities.

With respect to joint influence during the decision process, it is assumed that the routines which exist or are intended to be introduced in general are also to cover subsurface installations in overall planning.

On the whole, subsurface construction should be integrated in all sectors of overall planning. Physical planning, however, occupies a special position as the concrete co-ordinating planning for land use. Physical planning is also controlled by the most extensively developed legislation, the Swedish Building Law, which links planning with construction work. Subsurface installations are to be considered partly with respect to utilisation and partly in their capacity as buildings (which happen to be in the form of a rock cavity) from this angle, the criteria for application of the system of regulations in the Swedish Building Law governing utilisation of the subsurface and subsurface installations are satisfied. The legal interpretation of the building legislation currently in force does not, however, permit direct legal application to subsurface installations. A modification of the Law is necessary and more information in this respect is included in section 12.

The local authorities have a monopoly on planning. According to the Swedish Building Law the government and provincial authorities act as the inspection and ratifying bodies for community planning. Only in the case of certain installations which are important for the economical use of energy, wood fibre, raw materials and land and water resources is the matter given special consideration by the government.

This means that the local authoritity is the body which should be given the responsibility for initiating, planning and supervising utilisation of subsurface installations and subsurface construction work.

If the regulations and general applications of the Building Law are applied, subsurface projects covering a complete community as well as problems concerned with subsurface utilisation will be subjected to the administrative consideration which satisfies the demands made on an overall appraisal.

As was pointed out by way of introduction, the aim of planning for subsurface utilisation should be two-fold. On the one hand, the subsurface should be considered as a utility which requires more attention in total community development, while on the other hand there should be rectification of disturbance and damaging effects caused by deficient foresight in planning and construction work. In order to reach a radical change of attitude with respect to subsurface activities, it is not sufficient merely to rely on a wider range of training and more detailed information. For example the group of specialists and training units is small while the need for information is very extensive and covers not only those who come into contact with the problems, but also other groups who should be engaged in the utilisation of the subsurface.

It can be anticipated that the introduction of an administrative system of regulations will stimulate activity on the part of groups and bodies responsible for planning and decision to consider the problems associated with subsurface construction.

8.3.2 Detail planning

At many of the places which have been studied as part of the research project, subsurface installations have been considered and even included in the town plan. The reason for their consideration in the town plan has primarily been to ensure the right of disposition regarding the subsurface or to facilitate the construction of installations through another party's property. Further reasons are to protect installations against other forms of exploitation, in particular deep foundations, and to give the Building Committee influence concerning the location and design of the installations concerned. In the Stockholm city plan protective zones have been introduced which are intended to ensure the stability of the rock surrounding subsurface cavities.

In connection with urban planning, particularly of inner city areas, regulations have been introduced for two reasons to prevent the laying of deep foundations below a fixed level. One of these reasons is to prevent excessively deep building work and the other is to provide a margin for pure subsurface construction in the form of tunnels.

This means that the Building Committee is free to request the introduction of subsurface installations in the town plan and to incorporate in it preparation for subsurface construction.

It is a well-known fact from surveys carried out as part of the research project and from other sources that many subsurface installations are not described in town plans and that plans for them have not been subjected to examination in accordance with the regulations of the Swedish Building Law. There are many reasons for this. For example, there are no demands in the Swedish Building Law whereby subsurface installations are to be the subject of planning. In many cases, the Building Committee is not aware of the existence of installations because the owners of the installations have kept their activities covered by security regulations or have wished to limit the spread of information about these installations. When installations are built for the community and on land owned by the community, it has been considered that there was no need for urban planning. Since subsurface installations have not been interpreted as buildings, it has been possible to construct the installations concerned in areas of parkland without any planning consideration or building permits.

State and provincial authorities in Sweden are exempt from the obligation of applying for building permits but are under obligation to submit notification. Concerning subsurface installations, these institutions have not felt any reason to subject plans for them to local authority consideration and, furthermore, in many cases they have not even given the Building Committee notification of the project.

Recommendations

Since the town plan is an important instrument in planning for regulating conditions between neighbouring buildings and their effect on each other, and also to determining the location and conditions of construction it is obvious that subsurface installations should be included in the town plan.

It is less clear whether this should be necessary for installations in rural areas. As is the case with buildings constructed above surface, the procedure for examining the plan should, in rural cases, be considered in each individual case.

In the case of urban planning for buildings above surface, it is essential to specify how cavities are used, what the horizontal and vertical limits of the area of influence of the cavity are and also to indicate the position of openings in the ground. Furthermore, both protective and risk zones must be specified with respect to other subsurface installations or buildings above ground level.

In the description, it should be taken into account that subsurface installations require a greater degree of flexibility in practical application due to the possibility of structural deviations in the rock below ground level which cannot be anticipated. Due respect must also be taken to the need for future expansion.

In the survey material for the proposal for inclusion in the town plan, should also be reported the result of geo-surveys, consequence analysis with respect to influence on the surroundings during the construction and operation periods , as well as cost-benefit evaluation.

The degree of detail in the description is to be balanced against the type of installation and its position in relation to existing use of land and buildings as well as with respect to the risk of disturbance and damage indicated by consequence analysis, planning of the installation concerned and the town plan inquiry.

Building permits and control

In Stockholm, the metro stations are interpreted as constituting buildings but not the tunnels for the railway tracks. This means that the stations are to be subjected to examination for building permission. This has however only been the case with respect to

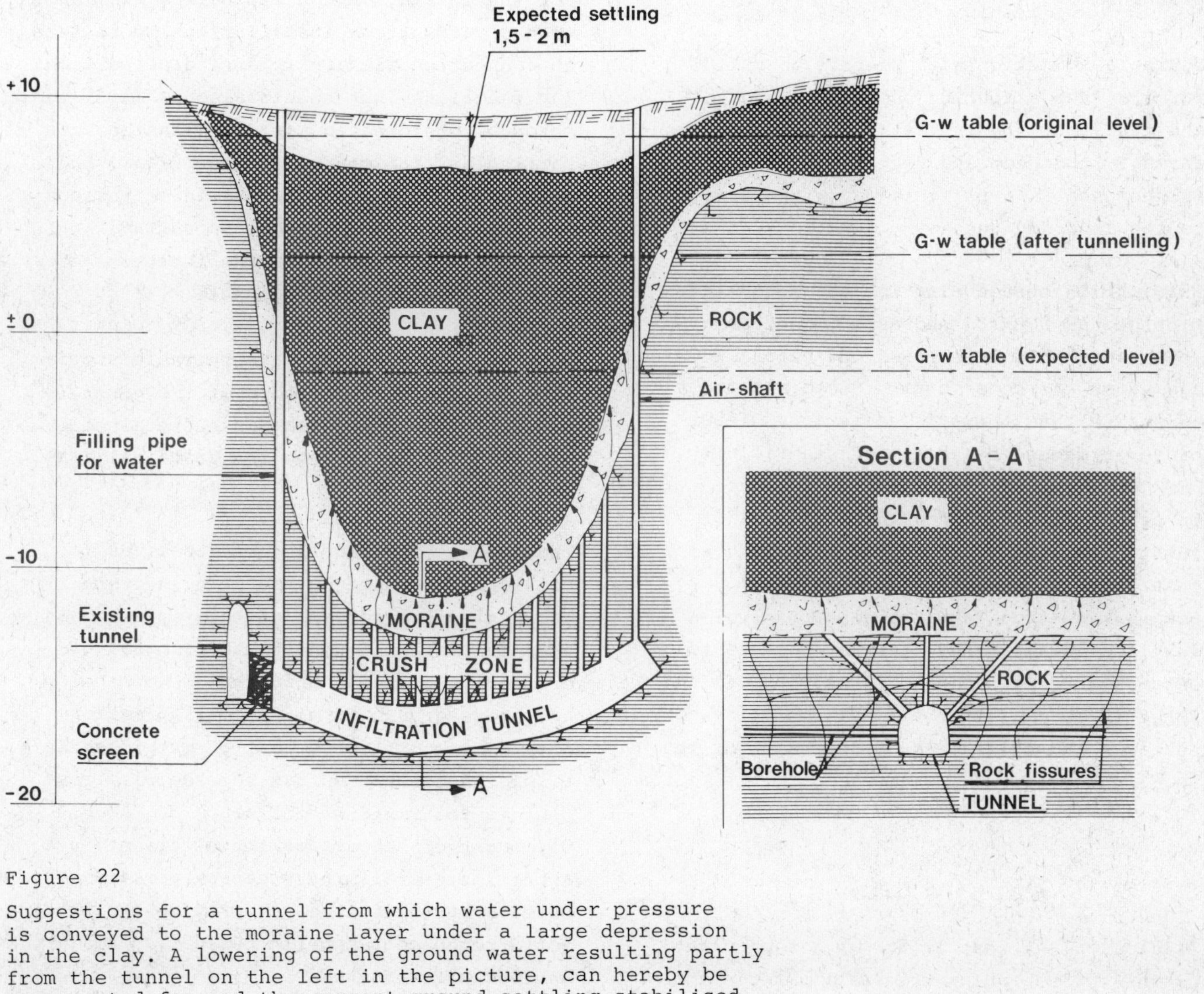

Figure 22

Suggestions for a tunnel from which water under pressure is conveyed to the moraine layer under a large depression in the clay. A lowering of the ground water resulting partly from the tunnel on the left in the picture, can hereby be compensated for and the current ground settling stabilised.

Stockholms VA-verk

ground level work at the stations. There are examples of the subjection to examination for building permission of subsurface installations in their entirety.

It appears important for subsurface installations to be interpreted as buildings for two reasons. In a large number of cases, they constitute places of work and environments where people spend time and therefore require careful design with public supervision. Secondly, since they are installations, they influence and are influenced by buildings and construction activities in the vicinity.

The need for building permits and control with respect to subsurface installations is an essential supplement to the demand for planning.

During examination with respect to building permits, due attention should be paid to the fact that the installation concerned conforms to the plan and that the installation is designed in a suitable way with respect to its intended use.

Examination should also include determination of the control and measurement programme which should be carried out before, during and after the construction stage, in order to ensure that there is no serious disturbance or damage to the immediate vicinity.

In principle, there should be control of subsurface installations corresponding to that for buildings above ground level. However, control requires expertise which very few local authorities have access to. It should be possible to satisfy the demand for this special expertise by employing consultants and by training personnel from the local authorities.

8.3.3 Partial plan

A detail plan, that is to say a town plan or the corresponding examination for planning, should be seen as an instrument for regulating the details for the construction of an individual projects or group of projects from the technical and legal points of view. One factor of even greater interest for overall appraisal is the planning that precedes detail planning, which includes alternative studies and a choice of alternatives. The concept of partial planning is intended to cover planning of a geographically limited area or a system of installations, for example, traffic arteries, service utility tunnels, etc. which are intended to be constructed from the medium-term perspective.

Partial planning assumes close co-operation with planning of particular functions with regard to, for example, supply systems of various types. It would appear convenient for the initiative concerning the examination of subsurface installations to be taken in connection with the planning of particular functions, but should also be initiated through co-ordinated partial planning. In particular, the co-ordination of, and the advantages of the co-ordination of various requirements, in multi-use subsurface installations should be in the interests of partial planning, and feasible.

Partial planning must make suitable provision for the participation of the general public in the consideration and evaluation of various alternatives for practical realisation.

The choice of subsurface alternatives for consideration must be based on an overall geo-analysis. Analysis of this type should also indicate the conditions for good co-ordination between construction work below and above ground level. Geo-information should be compiled so that a rough description of consequences can be made with respect to the risks of influence on the environment of subsurface installations. Approximate preliminary design work on subsurface installations can provide the basis for a rough cost-benefit evaluation of various alternatives.

Partial planning normally consists of co-

Cold store building at Årstadal; 1 service building, 2 entrance tunnel, 3 works and staff area, 4 transport shed, 5 stores, 6 refrigerated store, 7 locks, 8 cold store, 9 future extension.

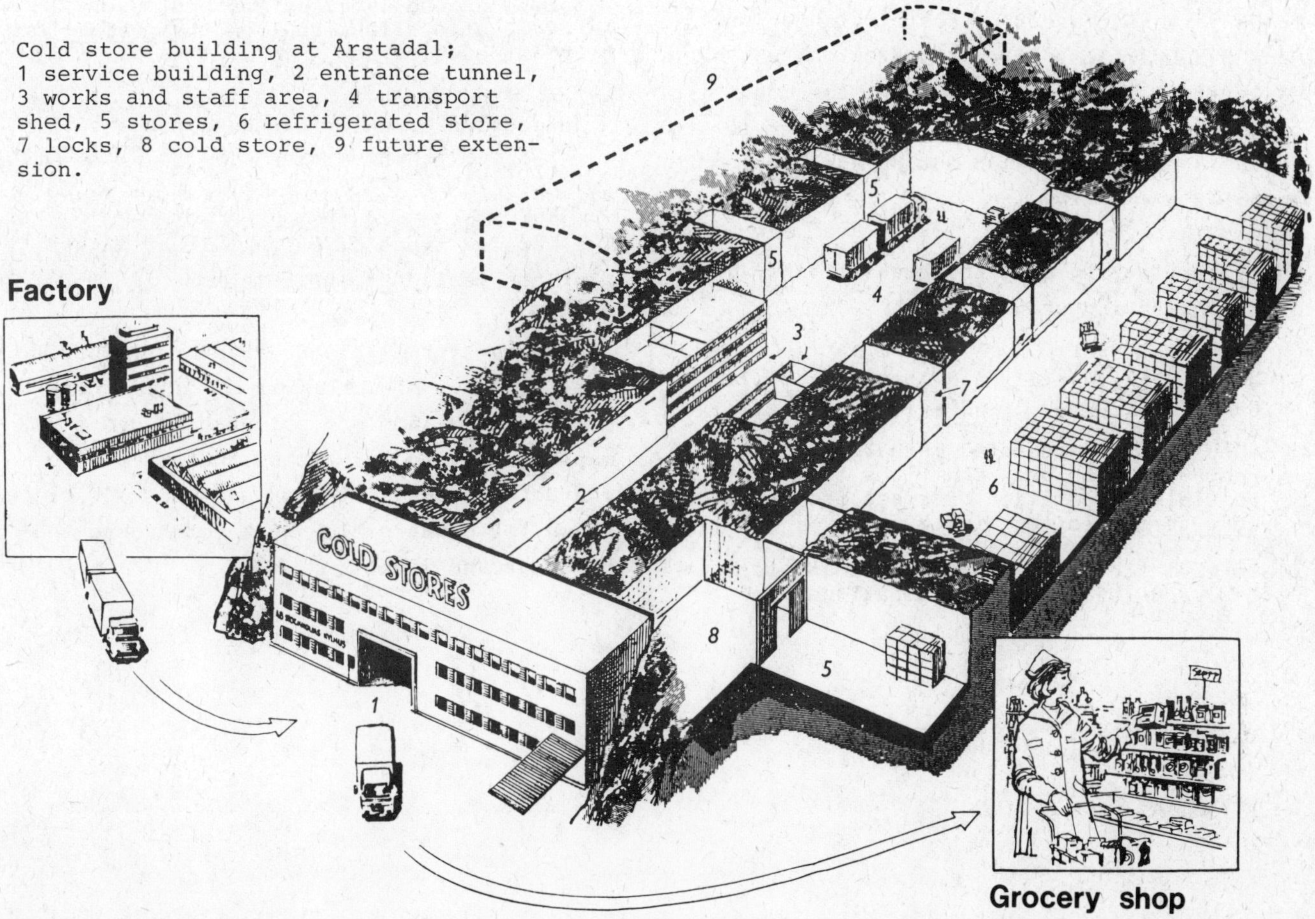

Figure 23

The cold storage installation for foodstuufs in a rock cavity located centrally in Stockholm. The installation costs are the same as for a similar cold storage building above ground, but the costs for refrigeration are lower. Because of the temperature inertia of the rock, not even a power cut of some length would be dangerous for the foodstuffs. The storage volume is 16 000 m^3.

Cold Stores AB

ordination planning carried out internally within a community, and with no statutory force. However, partial planning must result in a programme of proposal and action, in which the demands made by the planning of particular functions are satisfied and where the utilisation of the subsurface can be seen. Decisions concerning subsurface utilisation or the reserving of land for this can require the initiation of more long-term measurements with respect to settling and variations in ground-water level.

8.3.4 Other overall planning

The tendency to utilise subsurface installations stands in a relatively simple relationship to the topography and geology of the planning area and also to the structure, extent and density of urbanisation. There are thereby conditions for forecasting future demands on the subsurface even on the basis of overall formulation of goals and principles for future community development. With consideration to the fact that planning of buildings above ground level means various degrees of restriction on use of the subsurface, it is essential at an early stage to take into account the conditions governing subsurface construction in overall building carried out by the local authorities.

The co-ordinating and resource-limiting overall planning should

a) attempt to identify and locate probable subsurface needs in overall community development. Furthermore, these needs should be co-ordinated mutually and with respect to the geo-conditions to make possible reservation of space from the long-term point of view.

b) co-ordinate and balance the demand for resources of definable subsurface needs in time and space with respect to building above ground level at the best overall level of utilisation from the medium-term point of view.

c) describe subsurface projects, the execution of which has been prepared, in co-ordinated planning from the short-term point of view.

The various types of overall local authority plan are mentioned in Section 4.

All these overall plans can express the possibilities of utilising the subsurface of reserving land and orientating geo-mapping and geo-surveys so that the characteristics of the ground become better known for construction work both above and below ground level.

9. WHAT IS THE REQUIRED BASIS FOR PLANNING?

9.1 Registration

Various types of existing information make up an important basis for the planning and construction of subsurface installations. This concerns information about existing facilities, about reinforcement and sealing work in rock cavities, about existing deep foundation work and the results from rock and subsoil surveys carried out. If information of this type is to be easily accessible, it must be compiled and registered centrally with the local authority.

9.1.1 Facilities

The present situation

There are no requirements concerning documentation of rock cavities which have been constructed and consequently no demands on the registration of such data.

Certain local authorities, the metro in Stockholm, and others with major subsurface activities have carried out documentation of their own installations in a systematic way.

In not one single case has it been possible to find documented general information about all existing installations within any town. In certain smaller local authorities general information has been found in the form of personal knowledge, possibly passed on by word of mouth between those holding positions in the local authority concerned.

The situation in Stockholm and Gothenburg is completely different. Information on a large number of subsurface installations is dispersed over a considerable number of people holding official positions. It is impossible to form any overall picture of the total range of subsurface installations. Contributory factors include division of responsibility between various people holding official positions, varying quality of documentary material, etc. One illustration of the difficulty in obtaining a general view of the situation is the fact that in a recent inventory carried out in central Stockholm by the civil defence authority, air raid shelters with a total area of 3,000 m² were noted which had not previously been taken into consideration. Another clear illustration of the need for the registration of rock cavities can be derived from design work on the Stockholm metro. Rock cavities along proposed metro routes have been discovered purely by chance through the existence of ventilation devices etc., the result being a revision of the design work and other problems.

In a third example (Gothenburg Waterworks), a rock cavity was found to exist right in the path of a planned tunnel. The extent of the rock cavity was not documented by the person operating it and it was necessary to determine its dimensions.

Documentation concerning individual installations is generally to be found with the person responsible for it, but sometimes with the consultants or contractors employed. The quality of this documentation varies. Examples exist of very ambitious documentation, but also the opposite. In certain cases, this information has been found to be erroneous to a great extent (errors of several metres, which have resulted, in one known case, in a collision underground); in others, there is no documentation whatsoever. In many cases, the available documentation consists of working drawings; in other words, no measurements concerning the finished rock cavity have been carried out.

Desirable features

When planning and design work is being considered, there is an obvious need to know about rock cavities already in existence. This need applies not only to general information on the position of existing installations in the zone concerned, but also to the exact extent of existing rock cavities, both horizontally and vertically.

Within the mining industry, mines have been documented for more than 300 years. Maps are made up according to detailed regulations specified by mining laws and in the mining chart bulletins of the Board of Commerce. According to the result of the survey, including information from the organisation of mine inspectors, there can be no satisfactory plotting of rock cavities unless the same demands are made as in mining. Furthermore, the method used for describing installations is well known.

It is extremely important that completed cavities should be documented to their exact extent. There appear to be reasons for this being done according to the above-mentioned outlines.

The final document, today often referred to as the relationship document, should be formally registered with some authority in order to ensure its accessibility. With respect to the objectives of documentation, it is suggested that the authority concerned be the Building Committee. The suggestion put forward concerning consideration of building permits mentioned elsewhere in the survey (see section 10.3.3) is an indication in this direction.

The Building Committee should also be responsible for recording all registered subsurface installations in a system of general maps. This collection of maps must naturally be adapted to suit local conditions. In the opinion of the survey, an example of good service maps for complete towns should have a scale of 1:25,000, and the scale used in regions with many tunnels should be 1:4,000.

By way of comparison, it can be mentioned that installations above ground level are registered with the Building Committee in applications for building permits and in map collections. In a current survey, it is suggested that service lines be registered and updated in the service line maps and the register of service lines, the responsibility resting with the local authorities.

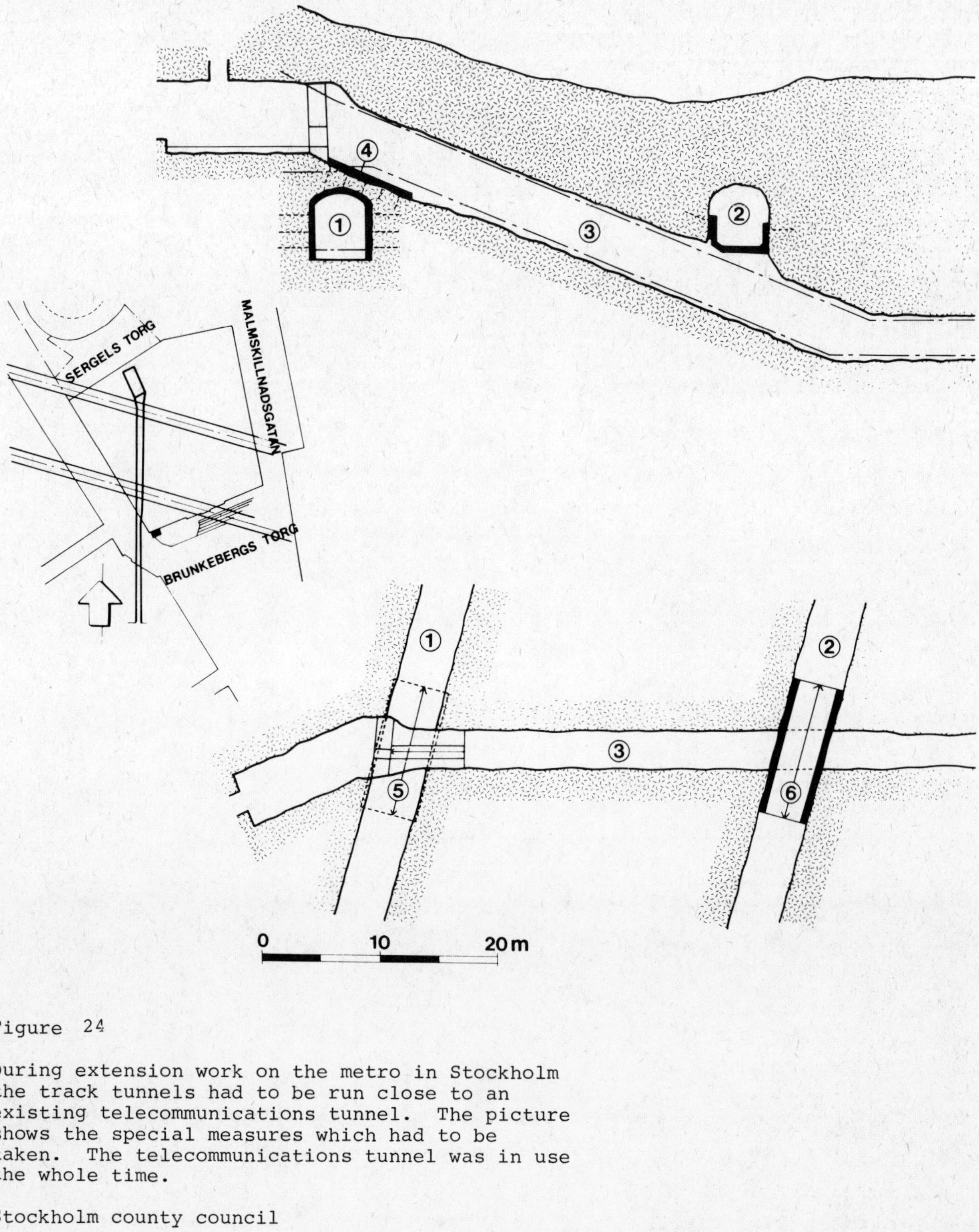

Figure 24

During extension work on the metro in Stockholm the track tunnels had to be run close to an existing telecommunications tunnel. The picture shows the special measures which had to be taken. The telecommunications tunnel was in use the whole time.

Stockholm county council

There is a strong argument for considering the co-ordination of different types of subsurface information.

9.1.2 Protective zones

In order to guarantee the constructional function of subsurface installations and in order to be able to use such installations for the purpose intended without interruptions, cavities in the rock must be surrounded by a certain protective zone. Within this zone, nothing may be done which can damage the installation concerned.

The intended function of the protective zone is to cover those zones subject to compressive pressure in the rock structure and should be normally designed so as to cover a region outside the limiting lines of the rock cavity defined by height, depth and la-

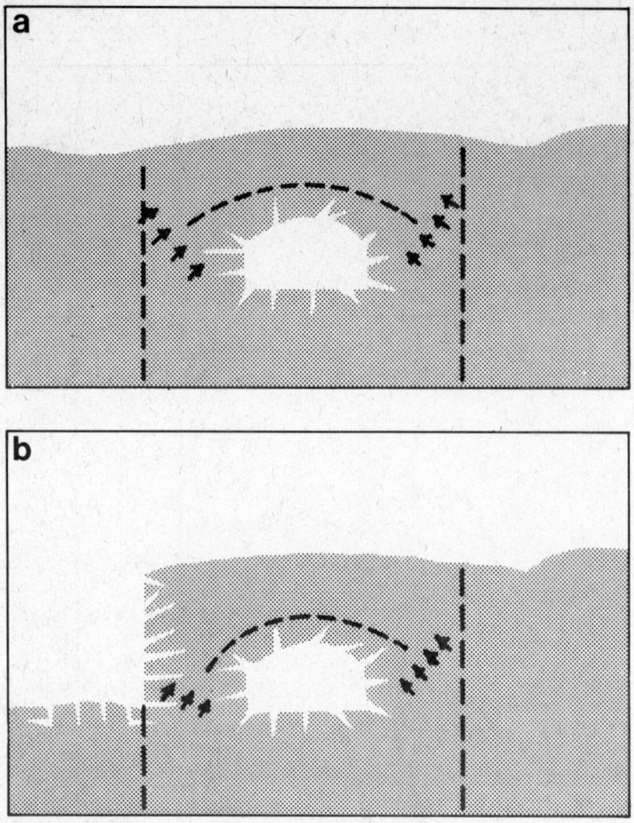

Fig. 25

A protective zone around an installation is meant to prevent encroachment on the bedrock within the zone to ensure continued existence of the installation. The size of the zone is often decided in relation to the cross-section of the subsurface cavity, without taking the local structure and extent of the bedrock into consideration.

Fig. a) shows a tunnel located near the surface. The protective zone is indicated with dashed lines. After excavation a natural vault with a pressure line and reactions has been formed.

In fig. b) ground blasting has been carried out in the vicinity of the tunnel. This results in increased cracking and increased deformation, which leads to poorer vault function and to the subsequent necessity of reinforcing the tunnel.

Dr. Sten G A Bergman

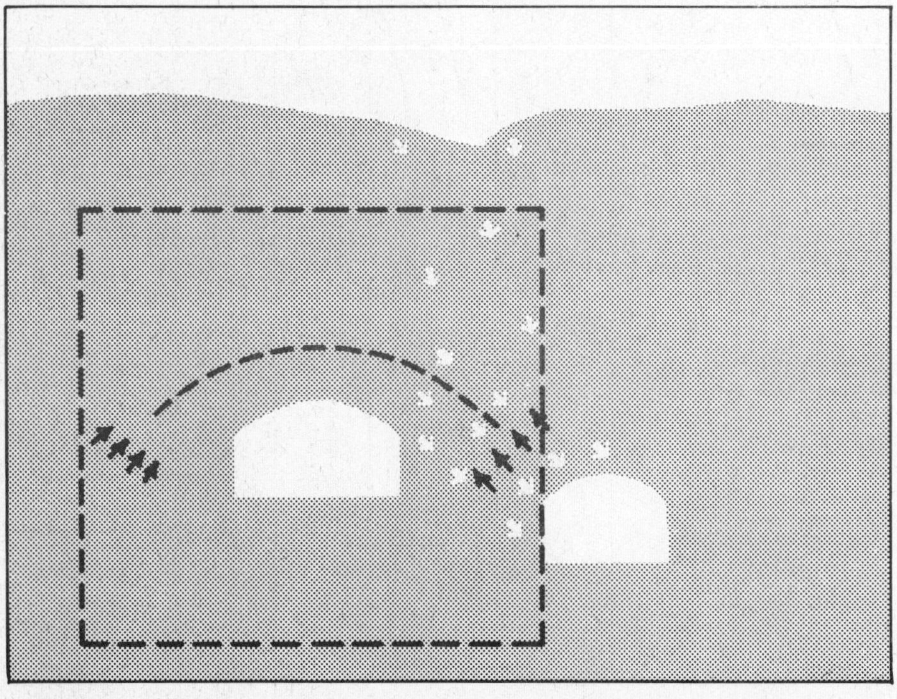

Fig. 26

A tunnel with a protective zone of the usual type. Beside the tunnel and outside the protective zone a new tunnel has been installed in such a position that the vault function of the first tunnel is weakened, with resulting large deformations.

Dr. Sten G A Bergman

teral dimensions, depending on the cross-section form of the installation, and on the demand made on safety. For example, in the case of oil storage facilities, protective measures must be taken to maintain the ground-water level. (A lowering in the ground-water level causes oil leakage).

As far as is known, protective zones round tunnels and rock cavities are only utilised regularly in connection with the metro in Stockholm. Other people responsible for subsurface installations have been directed to utilise zones of this type in areas which are considered to be particularly exposed, primarily by activities at ground level. In the case of most installations, however, no protective zone is specified in a way that can be maintained legally within the zone, which must, under all circumstances, be maintained intact for the continued existence of the facility concerned.

The extent of the protective zone is to be determined by the constructional conditions of the rock cavity, for example, the size of the rock covering, the structure and quality of the rock itself, the span of the rock cavity, etc.

The protective zone must always be marked on the main drawings of the installation. These drawings must also clearly indicate the extent of the protective zone both vertically and horizontally.

If a protective zone is to have legal effect, whereby it is respected in connection with construction work in the vicinity, then it must be specified in urban planning. The procedure is generally that the person in charge installation submits a suggestion concerning the extent of the protective zone to the person making up the plan. This person incorporates the zone in the town plan which is finally ratified by the Provincial Authority or the Swedish Government.

The outer limiting lines of the protective zone are marked in on the town plan as well as the level below which no subsurface work may be carried out. The town plan regulations specify what the different designations imply and other factors concering the planning zone.

9.1.3 The risk zone

Experience with respect to injury and damage and also risks of injury and damage has underlined the need of further protection around installations over and above that provided by the protective zone. Protection of this type, generally referred to as the risk zone, is being surveyed at the moment. As far as the metro in Stockholm is concerned, it is suggested that the zone cover ground level above the protective zone as well as a range within about 50 m on each side of the protective zone. Within this area, work such as excavation, sheet pile-driving, drilling, blasting, pile-driving or other forms of building work may only be carried out after application to and in joint consultation with the City Engineering Department.

The application of the risk zone is by no means obvious. Regulations in the town plan cannot have a purely legal basis and this implies practical problems with old town plans, which must be modified or revised.

In the case of the metro in Stockholm, it is suggested instead that a special system of maps be made up, on which the risk zone is marked. The intention is that the map be submitted to the administration authorities concerned both inside and outside the city, these must obtain a statement from the City Engineering Department in connection with building surveys in the risk zones. The Department is thereby provided with the possibility of carrying out supervision to avoid damage due to building work carried out close to the metro, but outside the actual protective zone.

From a legal standpoint, the idea is based on the regulations in the Swedish Land Laws whereby people, when utilising property belonging to themselves or others, must pay due respect to the environment (Chapter 3, § 1).

It would appear important that the risk zone instrument be developed and given final form in respect of both function and application.

9.1.4 The security problem

The present situation

One particular problem concerns information on secret installations. In a planning situation, the person responsible for the planning work needs to obtain knowledge about all installations in the area in which he is interested. At the same time, information on secret installations cannot be specified in the working documents for practical security reasons. There is no fixed method for the planner to carry out a survey if there is some "secret" rock cavity within his zone of interest. In order to determine the existence of such installations, the planner is obliged in principle to visit all the officials, about ten in number, who are today responsible for such installations. The degree of uncertainty includes the type of installation concerned and also the possible consequences of a conflict concerning location.

Within small communities, there is generally no knowledge available on "secret" urban rock cavities. However, examples to the contrary have been found.

The application of security regulations is by no means uniform. For example, companies and state authorities can apply security regulations which are not strictly associated with defence aspects. However, according to government department lawyers who have been consulted, legislation and legal practice constitute no obstacle to prevent the Building Committee, as an authority, from studying information covered by security regulations. However, in this connection, it is the authority concerned itself which determines whether security regulations are to be maintained or not. Cases have been noted where community planners have been denied information on the location and extent of rock cavities, and this has directly counteracted the objective of maintaining security regulations.

Desirable features

The problem of obtaining information on the existence of secret installations could be solved as follows:

Information on the existence of all rock cavities (without indicating their function) could be kept by the Building Committee,

Layout drawings of non-secret installations which acquire safety regulation status by association could also be kept there.

In connection with planning work of a general character, access is needed to the information compiled on secret (and non-secret) installations. Since this type of compiled information reaches such a high level of secrecy that it cannot be kept by the Building Committee for practical reasons, the information should at least be available at some regional authority. A suitable place for it is the Defence Unit of the Provincial Authority.

Where necessary, the person responsible for planning or design work should have access to information about the exact form of individual rock cavities.

9.1.5 The registration of geodata

Geodata here refers to geotechnical, geological and hydro-geological data.

The significance of geo-information for planning purposes and also for design and construction work is, as mentioned at several places in this report, of vital importance to subsurface activities.

Geo-information is characterised by the fact that:

- it can serve various purposes such as
  . planning both above and below ground level
  . design work both above and below ground level
  . consequence analysis
- it does not age
- it is produced by many interested parties
- on many occasions it is expensive or difficult to obtain
- old surveys may concern ground which has later been built on, that is to say, which has become difficult to gain access to or which has become inaccessible.

The mapping of rock cavities is of special interest for subsurface construction work. The information on rock structure provided by studies made in a rock cavity are far superior to all survey results obtained by surveys made from ground level.

In other words, there is an obvious interest in being able to co-ordinate geo-surveys and to facilitate the re-use of old survey results.

This question is by no means new. A survey has been carried out (BFR report R 70: 1973 Geotechnical databank) where suggestions are given as to the way in which such activities should be carried out.

The present situation

The filing of geodata is carried out in various ways in the local authorities with which this survey has been in contact. To a considerable extent, the repeated use of surveys used by the administrative authorities concerned are based on personal memory.

Within the local authority administrations of Gothenburg and Stockholm, efforts are being made to co-ordinate surveys carried out by different official bodies and to stimulate the repeated use of older surveys. A geologist has been specially engaged for this purpose in Gothenburg. In Stockholm, the results obtained within the community are filed in a joint manual archive.

Geo-surveys made outside the local authority administrations (and not commissioned by them) do not, as a rule, reach these local authorities.

Geological plotting of rock cavities has been noted by the survey in a few cases. This plotting has been carried out in different ways. Some is of high quality (the determining of dimensions, the making up of geological maps, the plotting on the corresponding maps of reinforcement work carried out); other plotting is of the diary note type. The excuses given in most cases for not carrying out measurement, making up geological maps or documenting rock cavities in other ways are "insufficient resources, shortage of time".

Desirable features

Here we would like to refer to the ideas referred to in the research project mentioned above and in other connections. Measures should be taken for the compilation and storage of geo-information, so that this can be used again. The aim should be uniform production of material. One suitable way of compiling information would appear to be to demand the registration of geo-information. To comply with suggestions made in other connections in the report, it is proposed that the information be made available at the Building Committee.

9.1.6 Mapping of excavations

If the best result is to be obtained, the geological cavity maps should be made up directly at the face during the excavation work before operational reinforcement and dirt constitute obstacles. For practical and organisational reasons, it may be difficult to

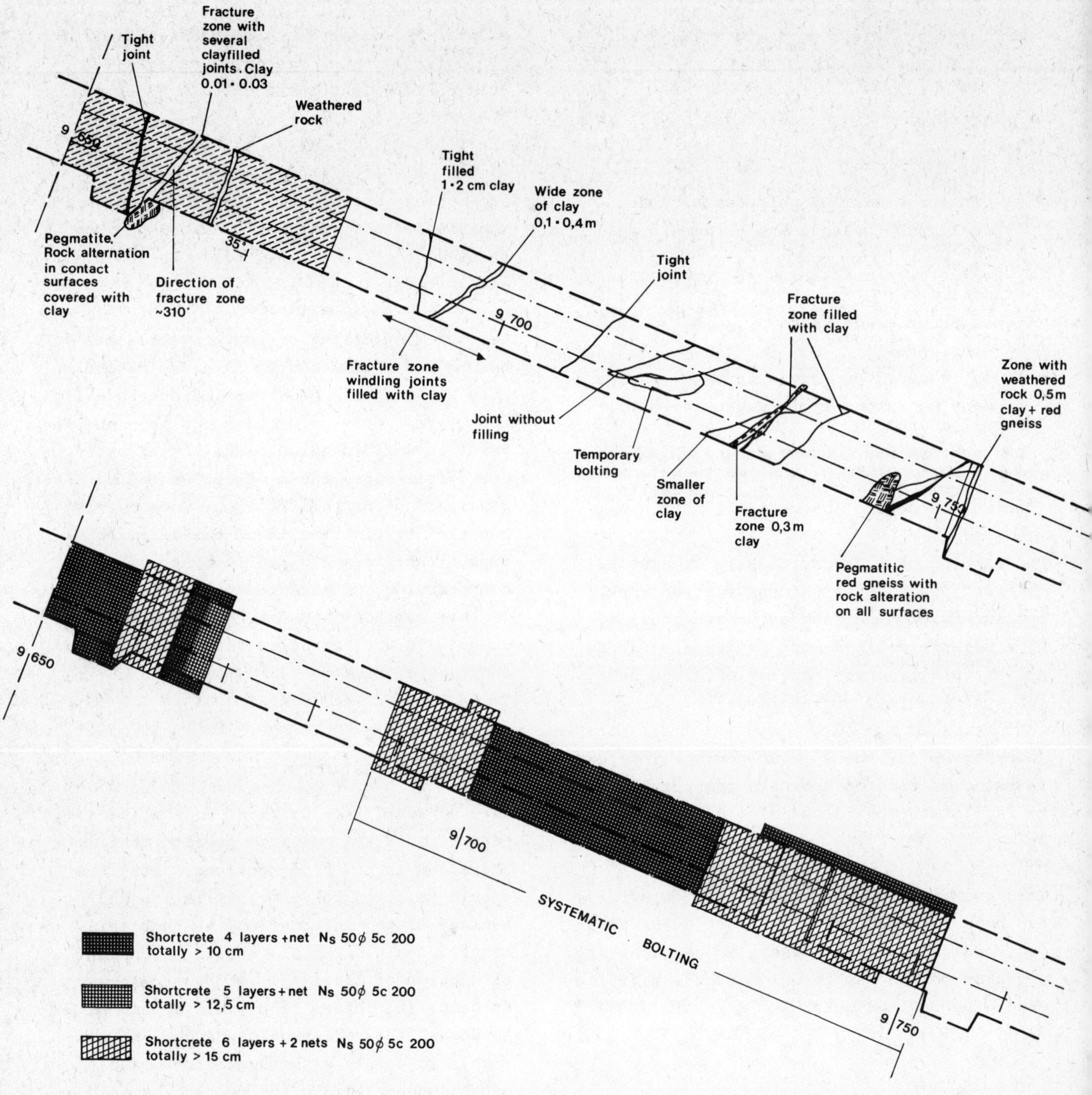

Figure 27

An example of geological mapping of a tunnel with corresponding measurement and mapping of support measures carried out. If anything unusual should be observed on a reinforced surface, for example, deposits of lime, then it can be immediately established on the geological map whether there is anything more serious in question. Where other subsurface installations are to be constructed in the vicinity, geological maps of tunnels are invaluable . Even comprehensive investigations from the surface cannot give such exhaustive information.

Gothenburg Public Service Department

allow a special geologist to carry out this work. It may be better to train a supervisor or someone else who is at the rock face in any case, so that he can carry out the work.

During the building stage, rock cavities are provided with operational reinforcement in order to protect human beings from injury and material from damage. The operational reinforcement is supplemented with permanent reinforcement in order to ensure that the subsurface installation lasts for a long time. The demands made on the stability of a rock cavity can be modified by both external and internal conditions, and unexpected events can occur. The rock can also change due to the fact that the ground-water in the vicinity of the rock cavity drains off and air obtains free access.

When supplementary reinforcement needs or other modifications are to be considered and when action is to be taken, the best conceivable basic material consists of maps and charts of the reinforcement measures carried out, in combination with geological maps. In other words, the charts made up of the reinforcement measures applied constitute important documents. Maximum usefulness would be achieved by making them consist of maps corresponding to the geological maps. The survey suggests that the plotting of reinforcement work be the responsibility of the person responsible for constructing the rock cavity and that the documents be registered in the same way as and together with the geological maps of the rock cavity concerned.

9.1.7 Reforms required

Subsurface installations should be documented in their true location when they have been completed. Documentation should be made up to meet the degree of accuracy which is generally used on maps and charts in mining.

The documentation should be registered with one single authority (the Building Committee is suggested). A set of maps should be made up so that it is possible to obtain a general impression of all the subsurface installations within the area of interest.

The handling of information concerned with rock cavities covered by security regulations should be revised. The administrative authorities should not complicate access to information which is of value for the planning and construction of subsurface installations. Security regulations do not appear to conflict with this. Secret information which, for security reasons, cannot be kept anywhere else, should at least be available at regional revel (the Defence Unit of the Provincial Authority is suggested). This applies particularly to the possibility of surveying in a central way as to whether there are "secret" subsurface activities within a certain area.

Regular surveys should be carried out of necessary protected zones and these should be legally specified in the detail plan. It is stressed that layout should be adapted to suit prevailing local ground conditions.

Risk zone documentation should be surveyed concerning needs, layout and application.

Geodata should be registered in order to facilitate and stimulate re-use. The particularly valuable making up of charts and maps concerning rock cavities should be included in this registration.

Maps and charts of reinforcement work carried out should also be associated with the geological maps and charts.

9.2 Appraisal of needs

Planning work includes an attempt to appraise the needs that must be satisfied. In the case of continuous planning activities of the land utilisation planning type, the appraisal must be graduated to cover various planning levels and various time phases.

The advantage of appraisals concerning the future can, particularly in the case of subsurface activities, be illustrated by means of several examples:

Knowledge concerning long-term requirements at national level:

- indications of the need for labour and training
- basic material for market analysis
- indication of requirements concerning development work
- basic material covering the extent of attention and activities by central and local authorities

Knowledge concerning long-term requirements at local level:

- providing the possibility of reserving subsurface accomodation in suitable positions
- providing the possibility of stimulating multi-use.
- providing the possibility of preventing conflicts and collisions

Knowledge concerning medium-term requirements at local level:

- providing the possibility of carrying out preparatory procedures in a suitable way and at a suitable time.

9.2.1 The present situation

It has not been possible to carry out systematic evaluation of arguments for or against subsurface location in a planning situation. On the other hand, several cases of ambitious analysis have been observed at the preliminary design stage whereby calculations of the cost-benefit type have been employed. Examples of this include the regional sewerage systems in Stockholm and Gothenburg, the Bolmen project, etc.

No compilation of basic knowledge has been made which includes analysis of the arguments for and against subsurface location. The information available is derived from analysis of individual cases or research studies of individual activities. The factors considered in this case have been limited to certain questions such as construction and operating costs, various technical aspects, and similar factors.

Statistical follow-up and descriptions directly aimed at subsurface construction in general and at any level are lacking.

Plotting and prediction at national level have been carried out in two connections. Before the OECD Conference in Washington in 1970, an investigation was carried out which, apart from matters concerned with technology, methods, research, etc., also included the present and anticipated extent of subsurface construction. On two occasions (in 1969 and 1974), Atlas Copco have carried out a survey of current and anticipated subsurface construction, mainly in the form of interviews, primarily with local government officials. (See section 3.3)

Initiative has been taken, however, to improve the general situation as regards information. The Swedish Council for Building Research has allocated a grant to the research project "Demand and Forecast", the aim of which is to provide basic material for the making of a general and continuous statistical description of subsurface installations.

9.2.2 Desirable features

It is possible to carry out an appraisal of the need for subsurface accommodation from a long-term point of view. Appraisal concerning the future is primarily based on the functional needs that may have to be satis-

Figure 28

When a sewage treatment plant is to be constructed or extended, location possibilities are often limited. Proximity to the recipient, the position of the existing sewer network, and the parts of older installations which can be used are decisive factors. As the possible location area is often more or less built-up, plants have, in many cases been sited underground for the sake of the environment (also often for economic reasons). A rather unattractive activity such as sewage treatment can in this way be effectively kept out of sight.

In both Stockholm and Gothenburg sewage is collected from larger areas via a comprehensive tunnel system to central plants. The picture shows such a regional plant, Henriksdalsverket in Stockholm, which serves about 700 000 people. The excavated volume amounts to 630 000 m^3. This plant is, therefore, the largest in the country. A housing area was built up on the surface of the rock at the same time as the plant was being constructed.

Stockholm public service department

fied. The individual functions must thereby be viewed from the perspective of collective community development. At a second stage, there will then be appraisal of the inclination to locate each individual function underground.

In spite of the difficulty of this task, long-term appraisals are still of the greatest significance, since these enable the preparation for co-ordination of different types of overall planning. This also indicates that appraisals should be carried out as far as possible with reference to basic planning material rather than to projection of trends.

The need for subsurface accommodation from a short-term point of view can be given in the form of a description of prepared projects, the implementation of which appears to be very probable.

Motives - factors exerting influence

Appraisal concerning the inclination to locate activities subsurface can be carried out with several different points of departure.

In general, the choice between a location above or below ground depends on how well the demands of the function concerned can be satisfied by one of these two sitings. The characteristic properties of the various sitings can be said to make up the motive for or against. Different types of external conditions - here referred to as influential factors - can be decisive with respect to the importance to be given to the various motives. In the final appraisal, several different motives must normally be considered together.

One effective way of comparing alternative projects is the application of cost-benefit analysis. The basic material for the analysis can be obtained by studying the relationship between the motives and the influential factors. It is assumed that these can be quantified and evaluated.

MOTIVES \ AFFECTING FACTORS	Topography	Geology	Urban structure	Social demands	Technique	Laws and adm.
Connection	①	X	X	X	X	X
Protection	X	②	X	X	X	X
Space	X	X	③	X	X	X
Environment	X	X	X	④	X	X
Cost	X	X	X	X	⑤	X
Time	X	X	X	X	X	⑥

Figure 29

Relationship between motives and influential factors

The figure shows a simple relationship diagram between motives and influential factors. In any given study, each "box" in the diagram can be given a value but further analysis is only justified when there is some significant difference between the alternative projects. Certain simple relationships between motives and influential factors are particularly prominent:

1. Complicated topography can well lead to the testing of communication lines in a subsurface position.
2. When protection of an installation is the target, subsurface location in good geological material provides advantages.
3. In large city areas where buildings are located close together, need for further accommodation often makes necessary the testing of supplementary accommodation below ground level.
4. The need to protect people and the natural environment from disturbing or unpleasant activities can result in their location below ground level.
5. The extremely advanced technology of subsurface construction can imply that constraction costs become lower than for a location above ground level.
6. Legal consideration of subsurface installations is simpler due to limited right of disposition and can therefore mean signeficant time saving.

Analysis methods

In a planning situation at a local or regional level, appraisal of subsurface needs from a long-term point of view can be carried out as follows.

The basis for appraisal is a forecast of the functional requirements that can possibly be located to the subsurface. Motives which are in favour of subsurface location in the various cases are indicated. The local conditions (that is to say influential factors) in the form of topography, geology, urban structure, and so on are identified in a general way - to the extent that they are already known. As a result of systematic evaluation, for example using the cost-benefit method, appraisal can be carried out concerning future requirement for subsurface accomodation.

The procedure described assumes a good level of knowledge of the advantages and disadvantages implied by subsurface location for various types of activity compared with alternative locations above ground level, as well as skill in evaluating the local conditions for subsurface construction. Knowledge of this kind has not as yet been compiled. A valuable step in this direction would be the creation of statistics describing existing installations, their functions and other characteristics.

General appraisal of the desire to build subsurface installations could also possibly be carried out according to the reverse principle. On the basis of analysis of existing and anticipated construction in a large number of places and of the local conditions for subsurface construction to be observed there, an impression can be obtained concerning the general inclination towards subsurface construction at a certain locality.

This reverse method has been used at national level in an American study. It is assumed hypothetically that demand for subsurface accommodation is regulated by such factors as increasingly cheaper construction techniques, increased density of urban built-up areas and efforts with respect to environmental conservation. It is assumed that subsurface demand is related to population density. This means, in other words, that forecasts concerning the demand for subsurface accomodation could be based on population forecasts. Other factors which should be considered are stated to be, for example, the continued influence of motor vehicles on urban structure and the form to be taken by environmental conservation efforts.

9.2.3 Proposal for systematic appraisal of subsurface location

For each type of activity that may be considered for subsurface location, appraisal is made as to how these special demands are met by location above and below ground. If a comparison is made between motives for location above ground and subsurface location, final evaluation can be carried out by considering only factors underlying the advantage or disadvantage of subsurface location.

The diagram describes the analysis procedure for individual types of activity. Demands on functional characteristics refer, for example, to motor vehicles being driven between two points at a certain speed, frequency etc., which, with given demands on vehicles and drivers, lead to requirements concerning range of vision, lighting, purity of air, load-bearing capacity, etc. This also includes restrictions on the resources which may be utilised for construction and operation. The structural demands are based on the functional demands and consist of demands on the physical characteristics of and equipment in the installation. In the motor vehicle example, this factor concerns height and width dimensions, radii, gradients, construction, separation from other activities, etc.

To the demands associated with activities exemplified above can be added the demands associated with a community. These are gen-

MOTIVES

ARGUMENTS IN FAVOUR OF SUBSURFACE LOCATION

ACTIVITIES	PROTECTION							SITE				ECONOMY		SAFETY			COMMUNICATIONS				ARGUMENT AGAINST SUBS. LOC.
	climate	damage	observ. from outside	disturb. "	" to environm.(operation)	" " " (inst.)	acts of war	suitable site	lack of space	freedom to exploit surface	separation	time-saving	cost	safety of operation	" , activity	" , community	short route	difference in level	water-course	other obstacle	how human beings experience subsurface
TRANSPORTS																					
fresh water		●				●		●	●	●	●	●	●	●	●	●	●	●	●	●	
storm water						●		●	●	●	●	●	●	●	●	●	●	●	●	●	
waste water				●	●	●		●	●	●	●	●	●	●	●	●	●	●	●	●	
hot water	●	●	●		●	●		●	●	●	●	●	●	●	●	●	●	●	●	●	
oil	●	●	●	●	●	●	●	●	●	●	●	●	●	●	●	●	●	●	●	●	
gas	●	●	●	●	●	●	●	●	●	●	●	●	●	●	●	●	●	●	●	●	
electricity		●	●	●	●	●	●	●	●	●	●	●	●	●	●	●	●	●	●	●	
telecommunications		●	●	●		●	●	●	●	●	●	●	●	●	●	●	●	●	●	●	
garbage	●		●		●	●		●	●	●	●	●	●	●	●	●	●	●	●	●	
waste, fluid	●		●		●			●	●	●	●	●	●	●	●	●	●	●	●	●	
waste, solid	●		●		●			●	●	●	●	●	●	●	●	●	●	●	●	●	
goods, bulk	●		●		●			●	●	●	●	●	●	●	●	●	●	●	●	●	
goods, part loads	●		●		●			●	●	●	●	●	●	●	●	●	●	●	●	●	
road traffic	●		●	●	●	●		●	●	●	●	●	●	●	●	●	●	●	●	●	●
rail traffic	●	●	●	●	●	●	●	●	●	●	●	●	●	●	●	●	●	●	●	●	●
TERMINALS																					
transport	●	●	●		●	●		●	●	●	●	●	●	●							●
telecommunications	●	●		●		●	●	●	●	●	●	●	●	●							
electricity (transform.)	●	●	●		●	●	●	●	●	●	●	●	●	●	●	●					
PRODUCTION (PROCESSING)																					
industry	●	●	●	●	●	●	●	●	●	●	●	●	●	●	●	●					●
power station		●	●		●	●	●	●	●	●	●	●	●	●	●	●					
water works	●	●	●		●	●		●	●	●	●	●	●	●							
sewage works	●		●		●	●		●	●	●	●	●	●	●							
STORAGE																					
oil	●	●	●		●	●	●	●	●	●	●	●	●	●	●	●					
gas	●	●	●	●	●	●	●	●	●	●	●	●	●	●	●	●					
hot water	●		●		●	●	●	●	●	●	●	●	●	●	●	●					
storm water	●		●	●		●		●	●	●	●	●	●	●		●					
waste water	●	●	●	●	●	●	●	●	●	●	●	●	●	●	●	●					
active storage	●	●	●	●		●	●	●	●	●	●	●	●	●	●						
passive storage	●	●	●	●		●	●	●	●	●	●	●	●		●						●
vehicle park	●		●		●	●		●	●	●	●	●	●								●
cold storage	●			●		●	●	●	●	●	●	●	●	●	●						
DEFENCE																					
sheltar		●				●	●	●	●	●	●	●	●	●	●						
command centre		●	●			●	●	●	●	●	●	●	●	●	●						
other		●	●			●	●	●	●	●	●	●	●	●	●	●					●

CHARACTERISTICS																	
NATURAL CHARACTERISTICS											USE						
SIZE/SHAPE						CLIMATE					STAY (IN SUBSURFACE)					METHOD OF INSTALLATION	OCCURRENCE
small	normal	large	protective zone	risk zone	imperviousness	dry (-wet)	hot (-cold)	ventilation	lightning	sound damping	short term/few	" " /many	" " /passage	long " /few	" " /many		
●											●						●
●	●										●						●
●	●				●			●			●						●
●							●				●						●
●							●	●			●						●
●							●	●			●						●
●						●					●						●
●						●	●				●						●
●						●	●	●			●						●
●						●		●			●						●
●	●					●		●			●						
●	●					●					●						
●	●					●					●						●
	●	●				●		●	●	●			●				●
	●	●				●		●					●				●
		●				●	●	●	●	●		●		●			●
	●	●				●	●	●	●		●						●
	●	●				●		●		●	●						●
		●			●	●	●	●	●	●				●	●		●
		●				●		●	●					●			●
		●												●			●
		●						●	●					●			●
		●			●			●			●						●
		●			●			●			●						●
		●									●	●					
	●	●									●						●
	●	●						●			●						●
		●				●	●	●	●		●						●
		●				●	●	●	●	●		●		●			●
		●				●	●	●	●	●		●		●			●
		●			●			●	●					●			●
		●			●	●	●	●	●	●		●					●
		●			●	●	●	●	●	●		●					●
		●			●	●	●	●	●	●		●		●			●

OUTLINE FOR CLASSIFICATION OF SUBSURFACE INSTALLATION

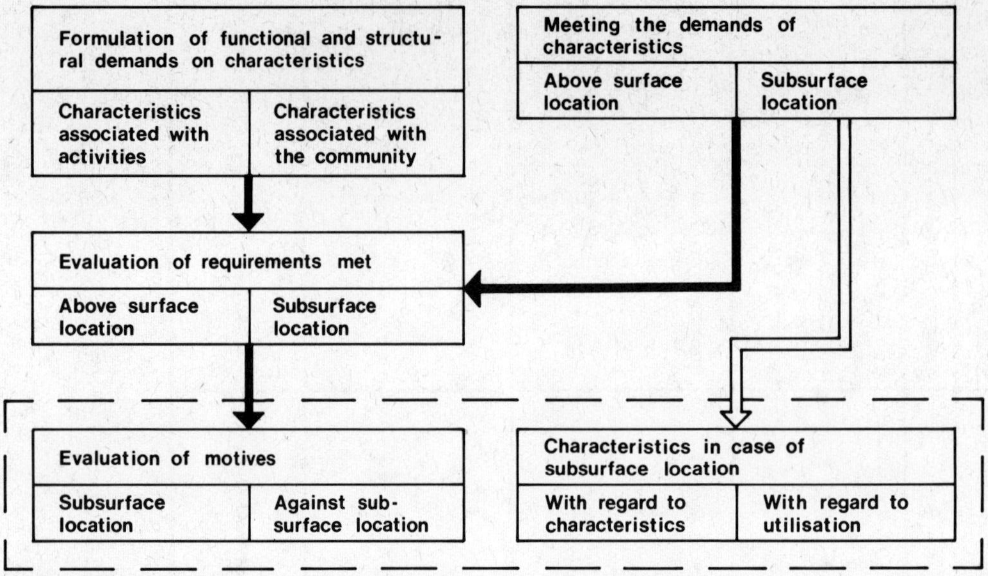

This part of the analysis is described in the classification diagram.

Figure 30

Proposal for systematic appraisal of subsurface location

erally not based on the activity in itself, but on other activities. They often consist of restrictions of an environmental type, such as demands on freedom from noise.

As an illustration of the analysis method outlined, there is a classification diagram consisting of the two lower "boxes" in the principle diagram above. The diagram is not complete and lacks, for example, media for expressing graduation or quantification.

9.2.4 Needs for reforms

The need for analysis is closely associated with planning activities and this means that the arguments concerning both these activities are practically identical. The same can be said about the need for a wider range of general information about subsurface activities. These arguments are therefore not repeated here.

Certain contributory factors of immediate value for the appraisal of requirements are important.

1) A systematic run-through of arguments for and against subsurface location of different types of activity. A corresponding run-through of factors which can have a certain influence on the choice of location. (Suggestions concerning the planning of such studies are given in the following section.)

2) Continuous statistical descriptions of subsurface construction are important. (See the research projects referred to above.)

3) A presentation of experience involved in the construction and operation of different types of subsurface installation could make up a valuable contribution towards improving the general situation as regards information.

9.3 Geo-planning

As has already been indicated, subsurface installations consist of cavities where the surrounding subsurface mass makes up the actual construction material. This means that planning and construction must necessarily be based on the appraisals from the viewpoint of construction geology. It must be possible to construct subsurface installations under safe circumstances and they must last for a long time. It must be possible to appraise beforehand and prevent the risk of undesirable effects on the environment.

9.3.1 The situation today

The extensive construction of subsurface installations being carried out today is at a high technical level. Experienced clients, design staff and contractors have co-operated in the development of utilisation methods, preliminary surveys for and construction of subsurface installations. Considerable efforts in the fields of research and method development have made it possible to reduce and to a great extent regulate the problems primarily concerning lowering of the ground-water level which came into the picture during the 1960's. Correctly used, the technology of today appears to correspond on the whole to the demands that can be made and which have been included in this report as examples.

At the same time, on the basis of surveys carried out, it must be stated that the implementation of many projects studied does not reach this desirable level. The criticism put forward below must be viewed against the background of what has been said above.

Urgency

Most subsurface projects are prepared and carried out during short and often during the shortest possible periods of time. As a result the various steps in the implementation of the project have not been carried out in a desirable way.

This can be illustrated by some observations from the areas surveyed:

- Time consuming preliminary surveys, for example, concerning the natural variations in ground-water level, have not been carried out over a desirable length of time.
- Detail design work on subsurface installations has been started before the results of rock surveys, on which it is based, have been available.
- Construction work has been started (at one end of a tunnel) before design work has been completed (at the other end).
- In the case of particularly urgent work, construction has been carried out without any design basis, since there has not been time to prepare this. Orientated data has had to be sufficient and the drawings have been delivered afterwards.
- There has "not been time" to document rock cavities, their geology, reinforcement methods, leakage quantities, etc.

Surveys with respect to the structure of the surrounding mass

In a surprisingly large number of cases, the standard of preliminary surveys with respect to the structure of the surrounding mass has been unsatisfactory. In the case of many tunnels with a small cross-section area, but also in the case of some major projects, the surveys have been limited to visual inspection of the material above the projected installation, a check of the position of the rock surface and a test of rock quality in the construction area by means of drilling. In a couple of the projects studied, no test drilling has been carried out at all.

In the case of many of the installations concerned, unexpected difficulties have been encountered as a result of insufficient knowledge of the structure of the surrounding material.

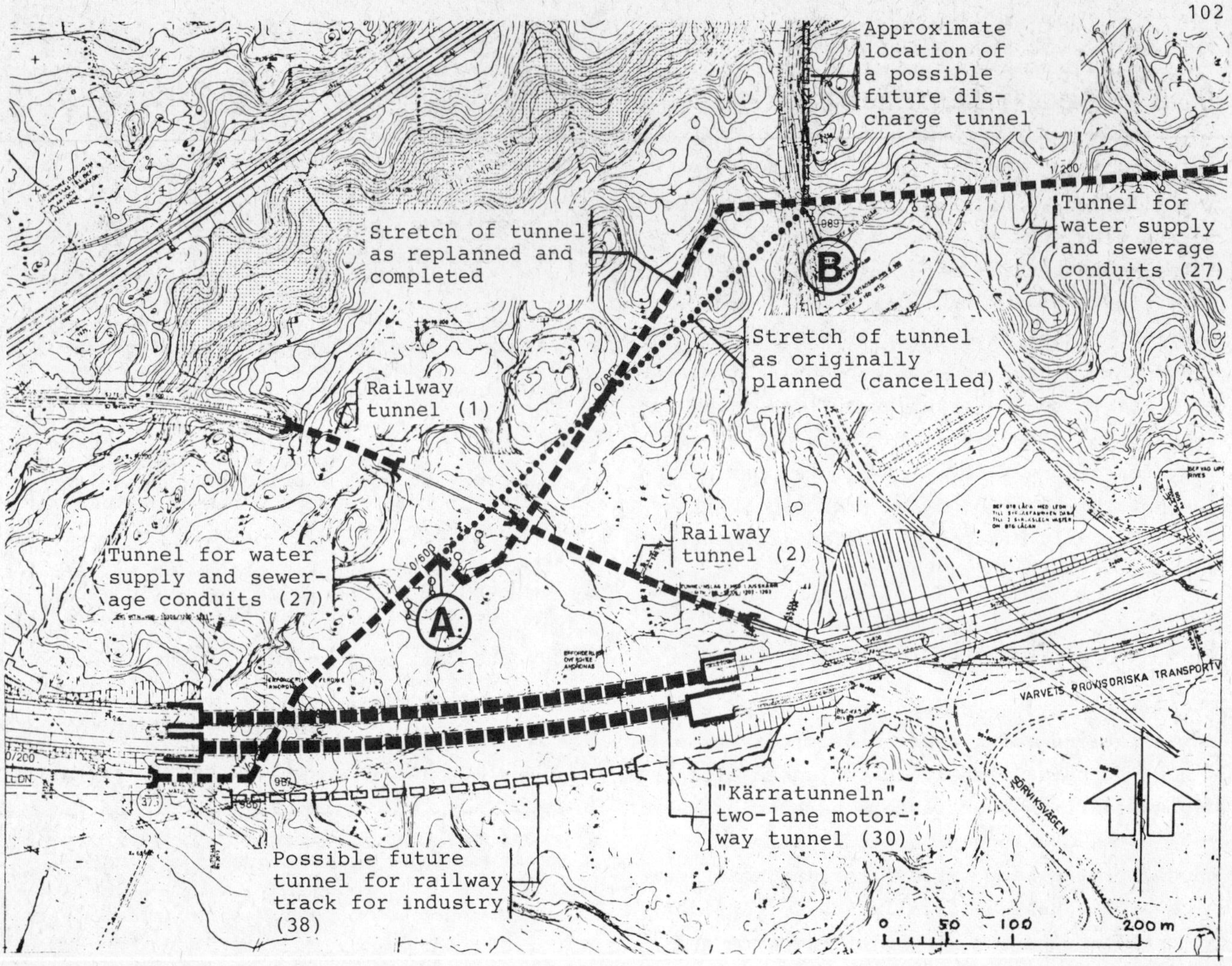

Figure 31

The planned stretch of a tunnel for water supply and sewerage conduits between points (A) and (B) was cancelled since a difficult crush zone was encountered during installation work. A new stretch of tunnel was worked out on completion of certain supplementary geological surveys.

VBB/Uddevalla Town Council Supervision office 1974

Examples:

- Rockfalls have occurred
- Great difficulties have been encountered in reaching firm rock at the initial blasting points.
- It has been necessary to modify the route of the tunnels during construction due to difficulties with poor rock quality.
- It has been necessary to turn installations relative to the crack system in the rock, this giving rise to revised design work.
- Unexpectedly poor rock quality has necessitated reinforcement work to an unanticipated extent (very usual).

Events of this kind mean delays in the project, risks to personnel and materials and - most often - considerable increases in costs, for example, because the labour force and machinery cannot be used as planned.

Surveys concerning effects on the environment

In the areas studied, the aspect which leaves most to be desired consists of surveys aimed at clarifying the risks which may be created by the complementation of a project to the surrounding ground and buildings. In many cases, no surveys have been carried out to determine influence on the ground-water level. In other words, it has not been possible to obtain any insight into general geohydrological conditions. The survey has not found one single case where analysis has been carried out in which the person responsible for construction of the subsurface installation has given direct classification of the risks that work in connection with the installation can give rise to.

A large number of cases are known, for example, through the press, where subsurface installations have given rise to damage to surrounding buildings as a result of ground-water leakage. Up to now in Sweden, tens of millions of Swedish crowns have been paid out in the form of compensation. Two of the cases with which the survey has come into contact are described below.

1) In an area for major new development, the exploitation was preceeded by such operations as thorough geological plotting with regard to building. A tunnel was under the areas for planned building which were most sensitive to settling. No special measures were taken during tunnelling to prevent ground-water from leaking in, for example in the form of pre-grouting. It has been found essential to carry out very extensive measures later on to limit the damage that this tunnel has contributed to.

2) Through the creation of a tunnel, a ground-water reservoir was given a new outlet. Water ran in the bottom of the tunnel which was cracked by the blasting. The water could therefore not be seen. The ground-water level was lowered over a large area and this gave rise to extensive settling of the ground. This area was scheduled for new buildings in the local authority's planning for land use.

Comment

Deficiencies in the quality of the various preliminary surveys have primarily concerned those clients and those responsible for design work with limited experience of subsurface operations. Many of the clients would appear to be badly informed particularly in connection with ground-water problems, a fact that has moreover been pointed out by the people concerned themselves. Insufficient cooperation and co-ordination between the various parties may possibly be explained to a certain extent by the fact that the professional groups concerned do not normally cooperate.

One general criticism put forward is that the geologists - and to an even greater extent hydro-geologists - are not permitted to participate sufficiently in the project. In order to save on costs, it occurs far too often that preventive measures of the type concerned with sealing against ground-water leakage through pre-grouting, are all too seldom carried out to a sufficient extent if indeed they are carried out at all. The reasons of economy that can be assumed to lie behind this have been all too often shown to be reasons of false economy.

9.3.2 The content of geo-planning

Some of the planning and design work concerned with subsurface construction has more and more frequently been referred to as geo-planning. The figure above indicates the various types of geological appraisal in connection with construction included when consideration is being given to a subsurface installation. Furthermore, the range of measures that can normally be considered in the case of varying ground conditions include 1) the actual construction of the rock cavity, 2) measures to prevent detrimental lowering of the ground-water level and 3) measures if damage should occur.

The next figure describes the various types of information compiled to provide the basis for geological appraisal in connection with construction at the design stage. As can be seen, a balance must be established between the various fields of competence, i.e. geology, rock mechanics and hydrogeology. The functional demands as well as the various community and environmental aspects must be taken into consideration.

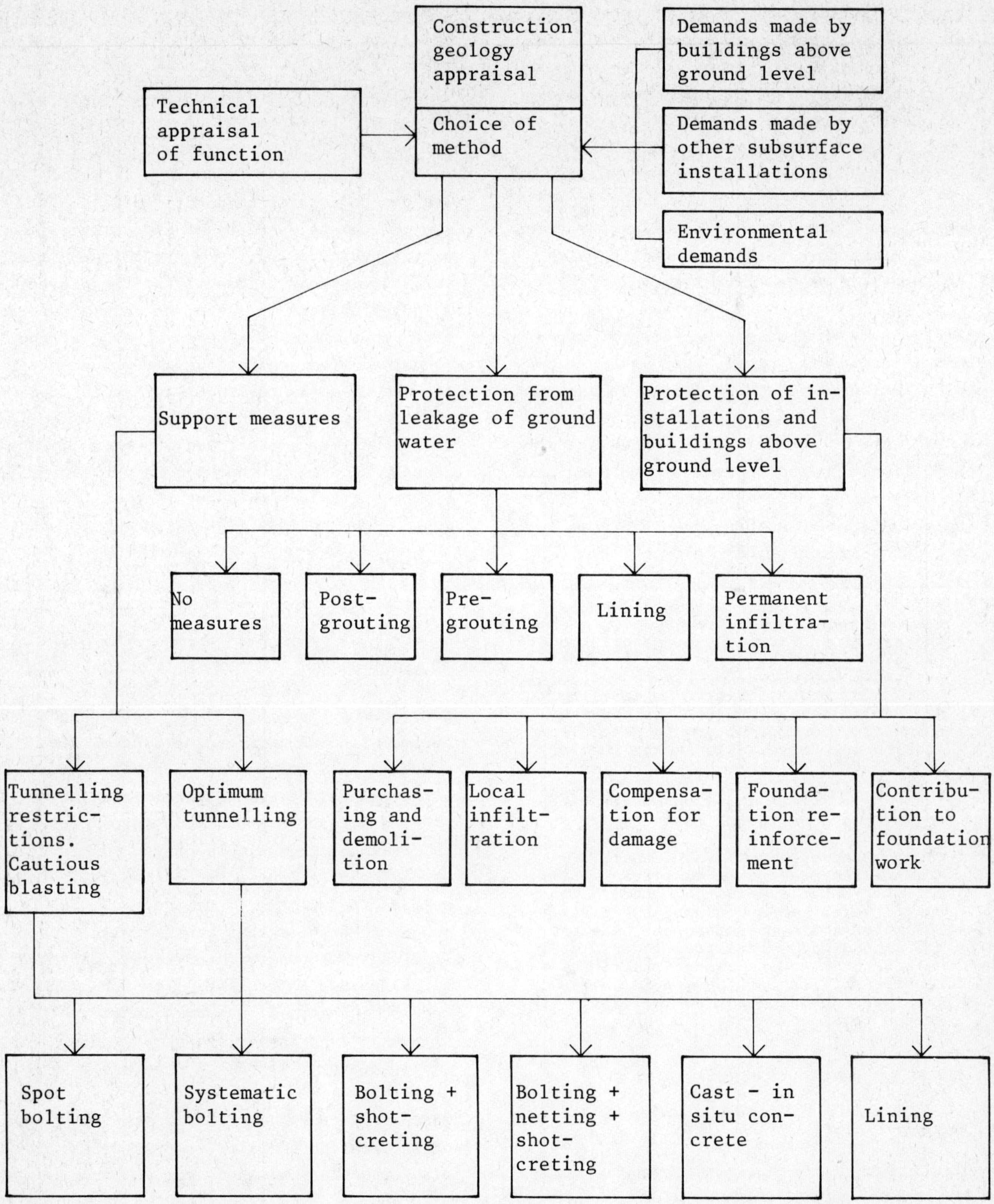

Figure 32

Functional diagram for subsurface installations in rock

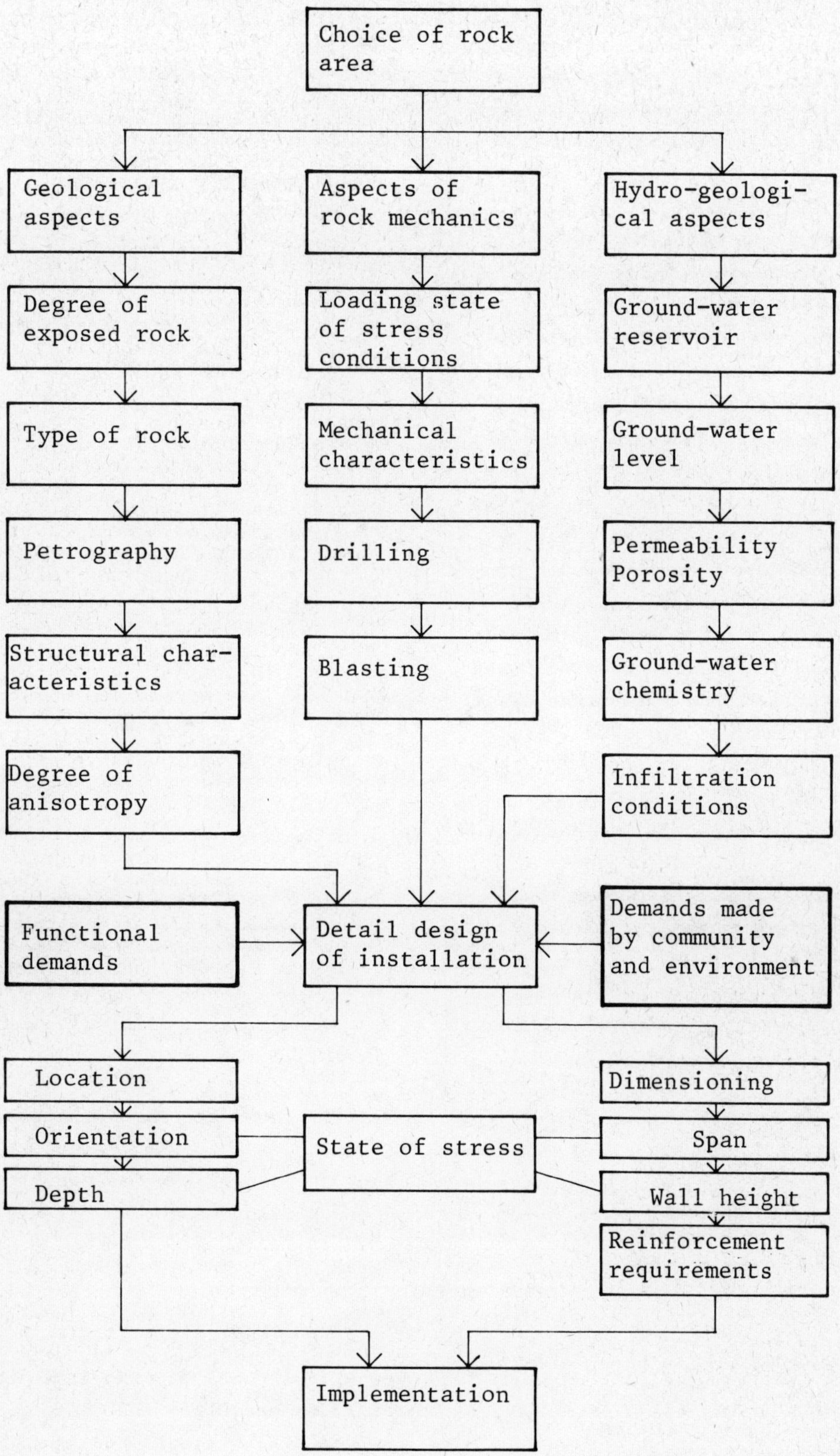

Figure 33

Analysis procedure in design work on subsurface installations in rock

Geology and rock mechanics (BeFo project 18)

The development of both knowledge and technology with respect to ground surveys has hitherto been split up to the extent whereby interest has been aimed at individual questions or development has been closely associated with companies. It has been difficult to produce any complete picture of survey technology and the relative value of individual methods. A current project within the outlines of BeFo financed by the Swedish Counsil for Building Research can be expected to be very important in this connection. It has the title "The value and extent of preliminary survey work - technical rock prospecting and documentation methods in connection with subsurface construction in rock". The aim of this project is to develop methods for technically and financially realistic and relevant forecasting which can form the basis for the design of different types of installation in rock.

In the first stage, an analysis has been carried out of nine subsurface projects which are going on or which have been completed, with respect to rock material and also to characteristics of and measures taken in rock construction technology, in order to master the difficulties arising. This information has been compared with forecasts concerning rock conditions at the depth concerned, based on geological appraisal and current survey methods. This stage is summarised in the following appraisal:

- Under certain conditions, preliminary surveys can give excellent forecasts of rock conditions.
- The forecast value of preliminary surveys is dependent to a great extent on the designer putting questions which are relevant with respect to the functional demands made on the installation. Some suitable form of check-list should be used as assistance when formulating such questions so that the information in the survey can be specified as early as during the planning stage. A model for a check-list of this type is being worked out.
- Satisfactory knowledge of the geology is a vital condition if the preliminary surveys are to be planned and designed in such a way that they provide the greatest possible amount of the information needed.
- Zones of weakness and other structural deviations are difficult to locate and specify with respect to character and extent at the depths concerned, and cannot always be indicated in connection with the geological survey of the ground.
- In order to carry out practical and applicable appraisal, the rock should suitably be subdivided into adjoining quality zones (= zones or areas of comparable quality from the viewpoint of rock construction techniques.) These quality zones can be limited and characterised by the use of various qualification systems for example Z (RQD) and Barton's Q-value.
- There will always be a certain demand for reinforcement over and above that which can be defined and quantified on the basis of preliminary surveys. There can be no technical or financial motivation for pursuing detailed information with expert survey methods in order to reveal this sporadic requirement for reinforcement.
- It is essential to clarify the geological factors which in fact give rise to requirements for reinforcement and also to be able to evaluate and define these so precisely that they are accessible when forecasts are being made. These factors must have technical relevance, be of a limited quantity, be definable during the preliminary survey and, from the perspective of scale, must match the problems in question.
- No objective appraisal can be carried out concerning final requirements for reinforcement in a subsurface installation without very detailed geological plotting of tunnel walls and roof. With respect to the fact that the subsurface and particularly the bedrock, can be considered from the practical and financial points of view as a facility which can in principal only be used once, such documentation should be compulsory.

73 different subsurface projects are to be followed up in the next stage. The results of this research work is expected to be presented during the autumn of 1978.

Geo-hydrology

Geo-hydrological mapping is intended primarily to establish the content of water in the material surrounding the rock cavity and the risk of leakage of water into the cavity. A secondary aim is to establish the connection between a suspected leakage of water into the rock cavity and the overlying ground water.

It is important, primarily in areas where there are buildings, to appraise the possible effect on the ground-water level or levels of a subsurface installation. The significance of this matter is described in more detail in the next section.

The appraisal of possible effects on the environment is based on knowledge concerning the ground-water flow, this being possible by studying different forms of flow pattern. In this way, difficulties concerning water inflow during the construction stage can be appraised, as well as the quantitative and qualitative aspects of the effect of the completed subsurface installation on the ground water. The appraisals are based on factors such as the chemical composition of the ground water, the content of tritrium, the temperature, and the various aspects of variations in the ground-water level (seasonal variation, air pressure, hydro-meteorological factors, tidal effects, etc.).

Surveys should be carried out over a long period of time, at least one year and possibly several, in order to be able to define seasonable variations.

9.3.3 The ground-water problem

Lowering of ground-water levels is one of the most difficult problem areas of all in connection with construction in built-up areas. This problem is characterised by many types of influential factors. The problem also applies to subsurface activities since tunnels and rock cavities can exert an influence.

Reasons for lowering of ground-water levels

In principal, four groups of reason can be differentiated:

1) Less possibility of natural infiltration as a result of urbanisation. A large part of the area in the form of buildings, road surfaces, etc. is hard-surfaced. The intensive character of precipitation makes necessary the conducting away of the storm water; otherwise the result would be flooding.
2) Drainage, that is to say the conducting away of the water, is made necessary by buildings which extend deeper into the ground than the natural upper level of the ground-water table. Action can consist of foundation work, service ducts, tunnels, etc.

Particularly in the case of tunnels and rock cavities, hydraulic communication occurs via crush zones and fissures, often in combination with water-carrying moraine on the rock surface, sometimes over long distances (several hundreds of metres). Communication to a point within an aquifer influences the whole of the aquifer to a more or less considerable extent (drawdown tunnel). Tunnels (extending long distances) can effect several aquifers.

3) The pumping of water from basements, cuttings for roads, etc.

   In the Water Law, the removal of ground water at a rate of up to 300 m^3/day at one and the same place is permitted. A permit is necessary from the Water Court for the removal of more water.
4) A natural influence can give rise to low ground-water levels. Examples of this are annual variations in precipitation, evaporation, air pressure. The current rise in the level of the land - 4-5 mm per year in central Sweden - is inevitable. In areas where there are no passage thresholds, there is a continual slow lowering of the ground water.

A temporary of permanent lowering of the ground-water level is generally the result of a combination of several associated factors. It is very rarely that one single factor, for example a tunnel, can be indicated. The presence of several factors involved in the lowering of the ground water implies, however, that one influence, in itself insignificant, can have a considerable effect since the margins with respect to acceptable lowering of the ground-water level have already been taken up.

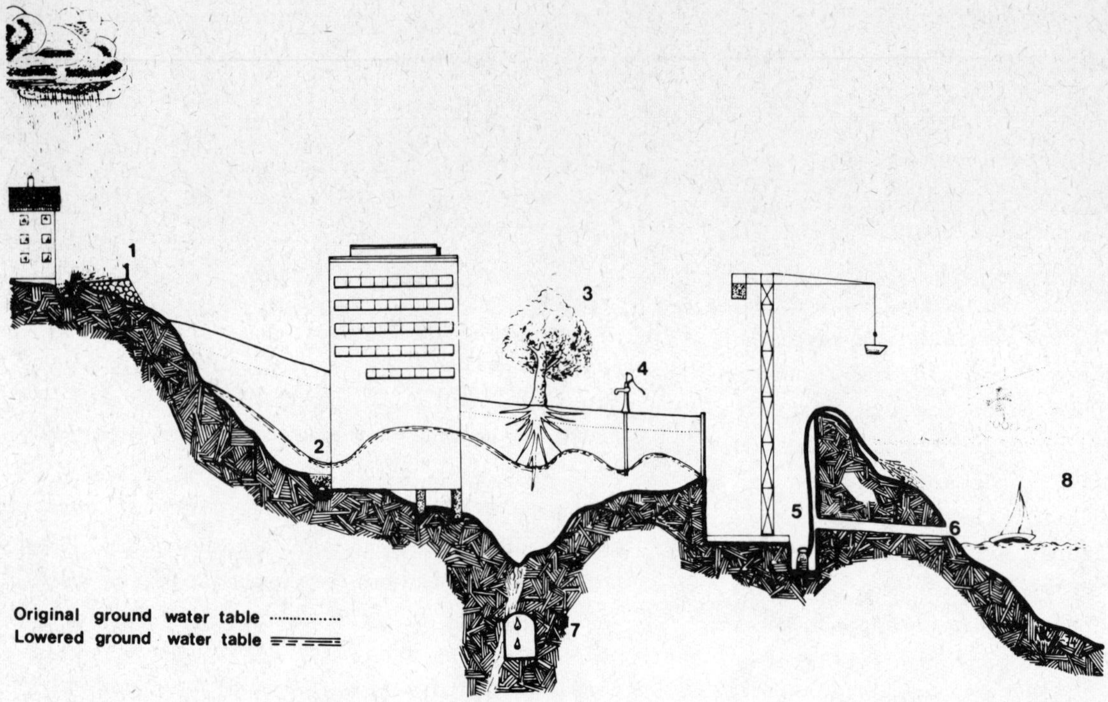

Figure 34

Illustration of factors contributing to ground-water drawdown:

1. reduced infiltration
2. underground conduits
3. vegetation
4. water outlet
5. deep construction
6. drawing off of ground water
7. drainage resulting from leakage subsurface installation in rock
8. raising of the level of land

Hagconsult AB

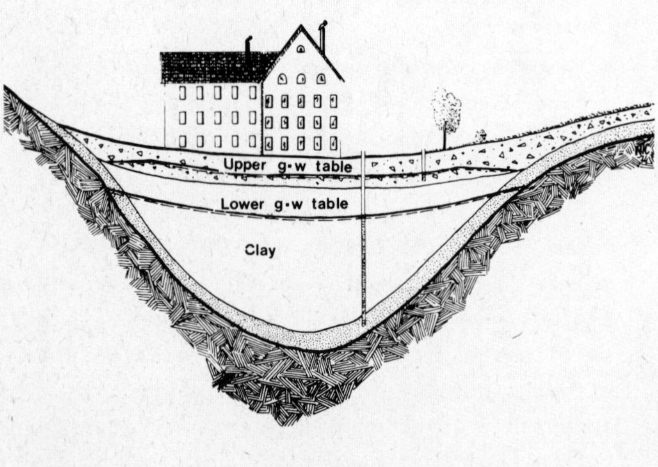

Figure 35

Example of layer sequence where two ground-water levels have formed, one in the fill above the solid clay and one in the moraine layer between the clay and the rock.

Hagconsult AB

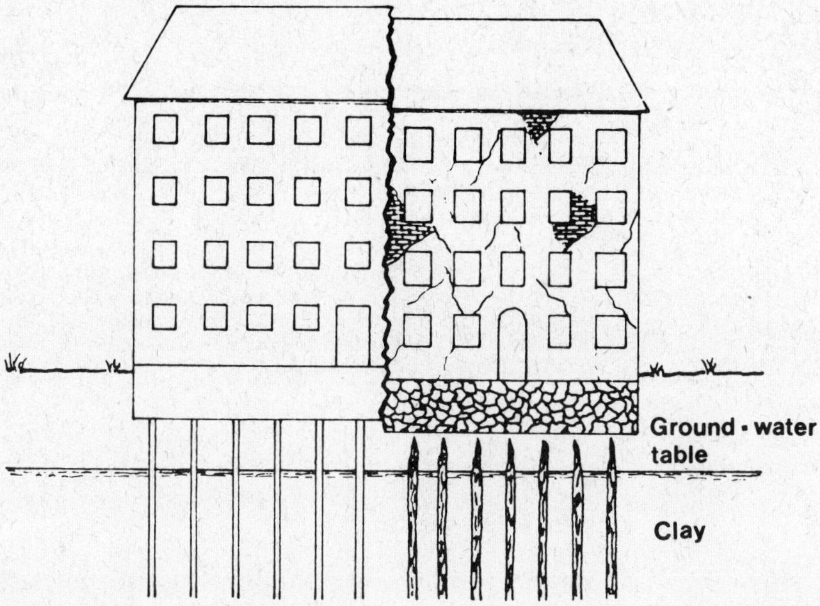

Figure 36

An illustration of a building with wooden pile foundations which is settling as a result of destruction of the piles.

Hagconsult AB

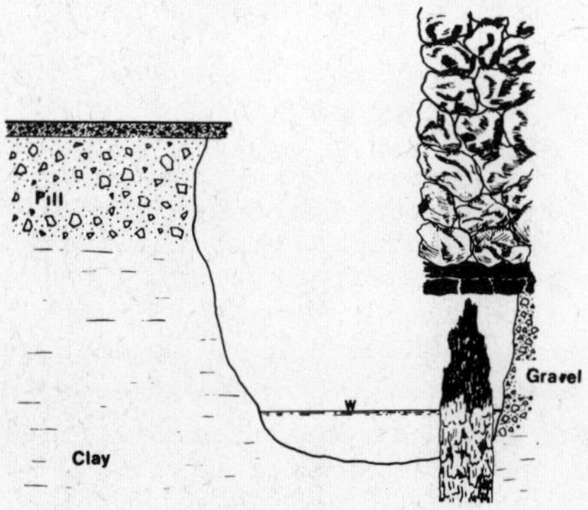

Figure 37

Detail showing a wooden pile which has rotted.

Hagconsult AB

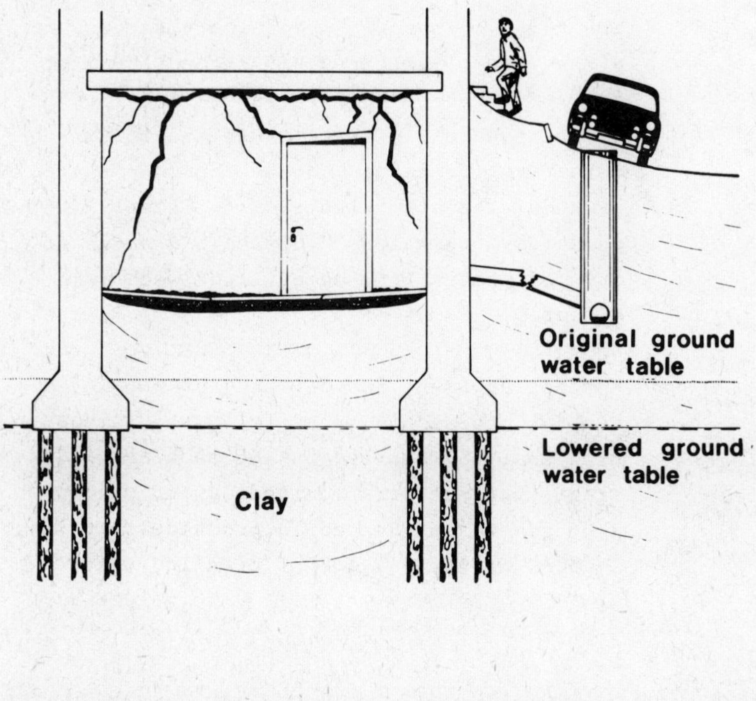

Figure 38

Damage resulting from ground settling
- cellar floor sinks and cracks
- partition walls break off
- conduits break off
- problems with staircase connections

Hagconsult AB

The consequences arising from a lowering of the ground-water level

- Settling of the ground
- Wells run dry
- The conditions for trees and plants are influenced, i.e. the ecological equilibrium is disturbed.

In this connection, interest is primarily directed at the influence on buildings.

- A lowering of the ground-water level means that foundations consisting of reinforcement beds or wooden piles come into contact with oxygen in the atmosphere and rot, in other words, they are destroyed. Damage of this type generally requires extensive foundation reinforcement work if the building can be saved at all.
- A lowering of ground-water pressure, primarily at the lower pressure level, means that the effective ground pressure in compressible ground strata (clay and organic earth strata) increases and settling of the ground develops. The extent of such settling depends on the thickness of the various strata, their degree of consolidation and also the extent of the decrease in ground-water pressure. In loose clay, settling can reach 1/10 of the strata thickness. Settling develops over a long period at a decreasing rate (10-100 years). The settling process can be checked by re-establishing the ground-water pressure, which it is however often impossible to do. (In many cases there are small amounts of water which maintain the pressure and therefore minor leakage can be decisive). The settling process, on the other hand, cannot be reversed.

The effect of settling on buildings can be of different types:

- In the case of uneven settling under a building, the frame breaks up.
- Buildings founded on piles do not sink, the result being a difference in height; pipelines and roads "rise", service lines to houses break when the land settles.
- Clay adheres to piles, thus increasing the loading on the piles. Such extra loading can be considerable even in the case of moderate settling and the piles can break.

The cost of rectifying damage

The city engineering authority in Stockholm has calculated some rough values for repairs to damaged buildings. (1974 price level):

- Foundation reinforcement of house in inner city area damaged by settling (five floors, area of each floor 450 m^2) about 0.9 million Sw.Crs or 2,100 Sw.Crs/m^2 surface. Added to this there may be repair costs which, in the case of "fairly serious damage" can cost about 250 - 400 Sw.Crs/m^2 surface area.
- The rectification of unacceptable settling round small houses varies between 10,000 Sw.Crs and 50,000 Sw.Crs. Basic reinforcement and repair of settling damage in the case of a detached family house costs about 75,000 Sw.Crs.

9.3.4 Geo-planning at different levels

The most suitable strategy in geo-planning naturally depends on the type of project and on the conditions regarding the ground and buildings in the area. It is, however, possible to make some general observations:

- Survey techniques are characterised by a very marked cost profile. General appraisal can usually be carried out with simple means and at low cost. With increasing demand for detailed information, the costs rise to a considerable extent. Surveys take a longer time to carry out and make keener demands on both personnel and equipment.
- The information which is required for building above surface goes a long way towards making overall appraisals of subsurface construction. The need for special supplementary information for subsurface construction is relatively small.
- Information about the structure of foundations does not age and therefore old survey results can be used again.

The value of general preliminary surveys has already been indicated. If a good level of knowledge can be acquired about local conditions at an early stage, then preliminary alternative locations can be chosen on suitable grounds. The effectivity of the costly detailed surveys can, therefore, with a greater degree of probability be utilised with more effect.

Another argument for devoting attention to general appraisals is as follows: geo-planning shall be carried out at such an early stage that the creation and layout of a project can be influenced in practice. It is at such an early stage of planning that

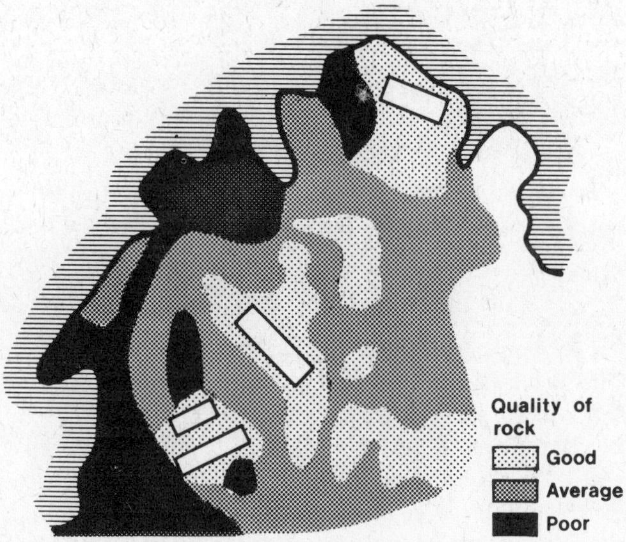

Figure 39

An example of an outline survey of qualitative grading of the bedrock and of how the location of rock cavities is done according to the quality of rock.

Hagconsult AB

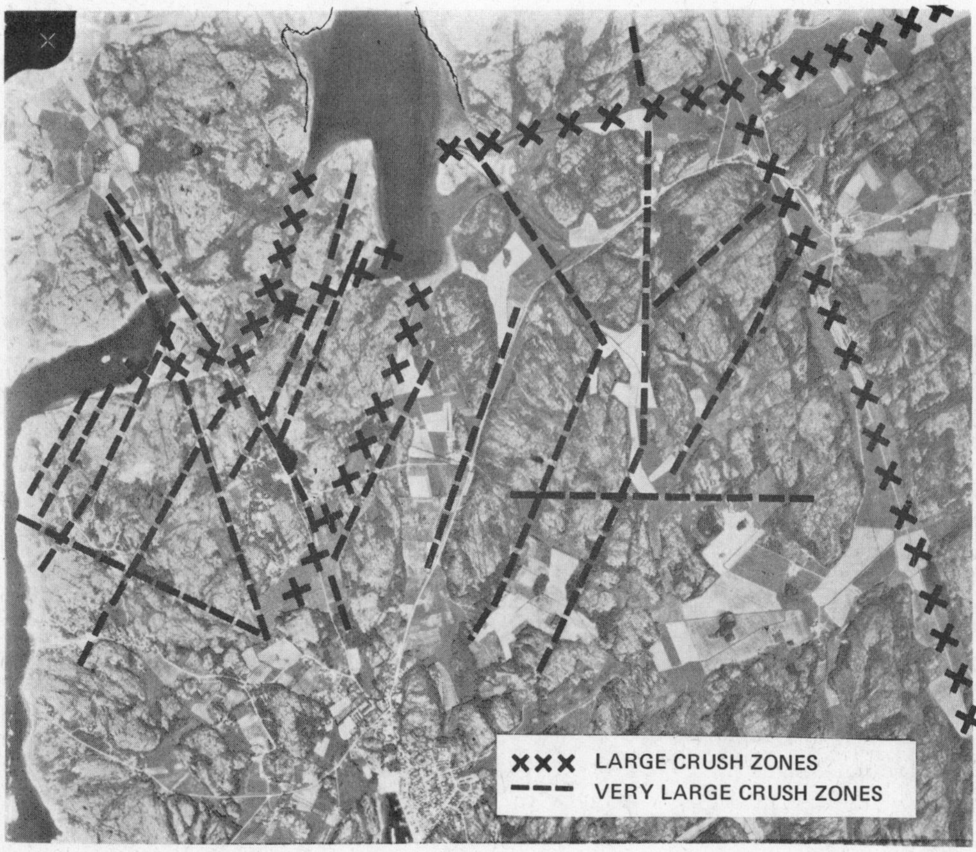

Figure 40

Example of remote sensing technique.
The interpretation of aerial photographs can as a rule give one a good idea as to the tectonic movements the bedrock has been subjected to and the resultant fissure and crush zones. The interpretation is prepared by studying all available geological map material. Aerial photographs of usual type are examined in a stereoscope by which method a visual model of the terrain can be studied. The interpretation should be calibrated by studies at the site. The cost is very low, for larger areas (at least 10 km^2) it costs normally about Sw Crs 200-400/km^2.

Vattenbyggnadsbyrån

troublesome problems and considerable supplementary costs later on can be reduced or avoided. If subsurface projects are integrated into community planning systems, there will be greater possibility of co-ordinating and making joint use of geo-surveys.

The geo-investigations which are initiated and carried out by various local authorities can be adapted to give information even for anticipated subsurface projects.

Measures at different planning stages

As already mentioned, in connection with overall planning it is possible to obtain a fairly good impression of the possibilities for construction by the use of simple and fairly inexpensive surveys. Unsuitable areas are rejected and subsequently avoided. Attention is paid to risk zones with respect to lowering of the ground water. Appraisals can be carried out if a subsurface installation runs the risk of weakening possible future buildings above ground level. An examination is made of the possibility of utilising existing geo-information or attempting to establish joint consultation with respect to survey work.

In partial planning, geo-planning is aimed at providing the basic material for decisions on the position and layout of one or more subsurface activities. Investigations shall be aimed at providing results which support appraisals of the economic and environmental assumption for subsurface projects. At this stage, decision should also be taken as to how the subsurface can in general be utilised to meet future needs. Surveys and long-term measurement concerning ground-water conditions should be initiated.

At the detail planning stage, geo-investigations are aimed at providing results which support decisions on implementation.

Geological, rock mechanical and hydro-geological conditions should be clarified on the whole. The conditions are specified for layout and construction work. Necessary protective measures are appraised on the basis of the anticipated risk level for consequences to the environment. Practical application planning should be prepared with respect to the operating organisation, the use of surplus earth and rock, etc.

9.3.5 The requirements concerning standards and instructions with reference to geo-planning and design

The demands made on the content and quality of the preliminary surveys mentioned in many parts of this report are aimed at the result which is to be attained. It appears important, however, for information to be issued as to how preliminary surveys and design work are to be carried out. Instructions of this type are available today for other types of building activity and are issued by the Swedish Board of Urban Planning. It is important to make up instructions for the subsurface field as well. Among the arguments for this are the following:

- There is a great need for information and advice as to the way in which subsurface installations should be prepared and constructed. Information in this respect is very limited today. The need for information has been emphasised in many contexts.
- These instructions make up a natural starting point when formulating the demands to be made in the suggested consideration and control proposed in connection with building permits.
- The instructions would provide valuable reference in legal appraisal of responsibility in connection with subsurface construction.

When instructions are being made up, demands are made on adaptability to suit varying types of condition. For example, the need for preliminary surveys is considerably less if it is intended to build a subsurface installation above the highest level of ground water than would be the case at a greater depth.

As an illustration of the ideas presented, there follows a suggestion as to how a recommendation should be formulated:

The recommended minimum level of preliminary surveys with respect to the building characteristics of the ground. Supply tunnel, cross-section area up to 50 m^2.

Well-stripped bedrock

At least one core drill hole per kilometre.

At least two earth-rock soundings in earth-covered sections when width exceeds 30 m.

Earth-covered bedrock moraine

Seismic profile (continuous)

At least 2 core drill holes per kilometre of tunnel

At least 10 rock-earth soundings per kilometre when rock cover is 30 m according to the seismic profile.

At least 20 earth-rock soundings per kilometre when rock cover is 20 m according to the seismic profile.

At least 50 earth-rock soundings per kilometre when the rock cover is 10 m according to the seismic profiles.

Earth-covered bedrock (gravel ridges - fault areas)

Seismic profile (continuous)

At least 4 core drill holes per kilometre of tunnel

At least 100 earth-rock soundings per kilometre where rock cover is 20 m according to the seismic profile.

At least 200 rock-earth soundings per kilometre where rock cover is 10 m according to the seismic profile.

The instructions should be supplemented in stages to cover all sections of preliminary survey and design work. There are needs of particular importance applying to basic material for appraisal of questions of responsibility. For the layout of rock cavities, there must be dimensioning standards for roof and wall sizes based on various rock structure parameters.

9.3.6 The part played by the local authority in geo-planning

The local authority departments can and should take various measures which facilitate geo-planning for subsurface installations. The local authority can, through its administration and committees play a part in collecting geo-information and making it available. Either by registering the investigations carried out or by processing the investigations into maps of built-up and anticipated subsurface utilisation areas. See also section
In connection with planning and other forms of work, geological and geo-technical information is compiled in the form of maps of different types which are clearly valuable aids for above ground as well as subsurface planning.

Of considerable general value are continuous level measurements which should be carried out at a loose network of points throughout the built-up sections of the community. Based on these reference points, local changes in ground-water level can be differentiated in a more reliable way.

Acting locally as an authority, the Building Committee is responsible for both new construction and renewal. The Committee sees that the current ground settling processes are followed up. In various communities routine registration of lowering of the ground-water level and measurements of settling already takesplace. In the centre of Stockholm, building foundation plans have been made up as basic material for the planning of renewal. These plans show the foundations of existing buildings, basement floor levels and ground-water channels, together with the results of measurements of settling. This type of information is also of the greatest value in the planning of subsurface installations, and should be made generally accessible.

The problem associated with a lowering of the ground-water level, described in section 9.3.3, causes considerable trouble in many communities. The possibility for the community to

Husgrundläggningskarta

Figure 41

To facilitate assessing the risks of damage in connection with a possible lowering of the ground water, maps of building foundations have been made up during the design of metro track sections which have recently been completed. The maps indicate the houses situated in the vicinity of the tracks, the method of foundation and certain details about foundations, such as pile material, pile cut-off levels, occurrence of mattresses, with levels, as well as types of basement floor (on ground and floating), with levels. In this way an idea can be obtained of what risks there are of damage if there is a lowering of the ground water or if settling occurs. The ground-water levels are kept under continual observation at several points along the track, before, during as well as after the work has been carried out. An account of all investigations is given in a geological and/or geo-technical report.

Stockholm County Council

act is limited since the Swedish Water Law permits the ground water to be removed, for example, by pumping out basements. Pumping close to the limiting value specified in the law, 300 m^3/ 24 hours in any one place, can have drastic consequences in particularly delicate situations.

It would be desirable if local authorities in areas which are sensitive to lowering of the ground-water level could formulate a hydrogeological target specifying acceptable decrease levels. In extremely exposed areas, for example, those with valuable older buildings on wooden foundations, extensive measures may be justified. A procedure of this type presupposes, however, that the local authorities are provided with the possibility of preventing detrimental removal of ground water to an extent of even less than 300 m^3/ 24 hours.

9.3.7 The content of a consequence analysis

When planning objects, in proposed urban planning and in the consideration of proposed building permits, an appraisal of the risks of influence on the environment should be included. Greater demands are made on accuracy in the basis for an assessment as the detailing of the project proceeds.

The problems involved in the lowering of ground-water level have already been described (see section 9.3.3) Appraisal of the risks of the lowering of ground-water level and ground settling must be related to the consequences which may result.

In unspoilt, undeveloped country, there are normally no detrimental effects. Nor does ground settling in built-up areas necessarily cause damage even though this often is the case. Within risk zones, a survey is made of buildings and installations with respect to type of foundation and general condition. Preventive measures may consist of pre-grouting sealing lining, infiltration of the water into the ground, etc. In sensitive positions, reinforcement of buildings can be considered, as well as the alternative of purchasing and demolishing the building.

The risk of damage caused by shaking and vibration from blasting work is studied in a comparable way. Where there is risk of damage, this is prevented by the use of a blasting method referred to in Swedish as cautions blasting. In special cases, full-faced drilling may be advisable. In particularly sensitive positions, the purchasing of buildings may be considered.

Initial blasting and other work which concers the surface of the ground should be planned and carried out bearing in mind the encrouchment disturbance on the environment and the surrounding scenery. This applies to transport areas and other arrangements which are only used during the construction stage, as well as to permanent facilities.

The removal and outgoing transport of earth and rock necessitate heavy and disturbing transport which can go on for long periods in the case of major projects. Outgoing transport should be planned so that people living in the vicinity are not subjected to severe disturbance.

9.3.8 The handling and disposal of surplus excavated material

The large quantities of earth and rock which are often excavated in connection with subsurface construction work should be used in a planned way. Within certain local authorities, special consideration has been given to this task. However, in cases the problem is today solved at the very last minute.

Surplus material consists on the one hand of valuable raw material and on the other hand of waste material which is problematic from the transport and disposal angle. If the surplus material can be utilised within reasonable transport distance, it

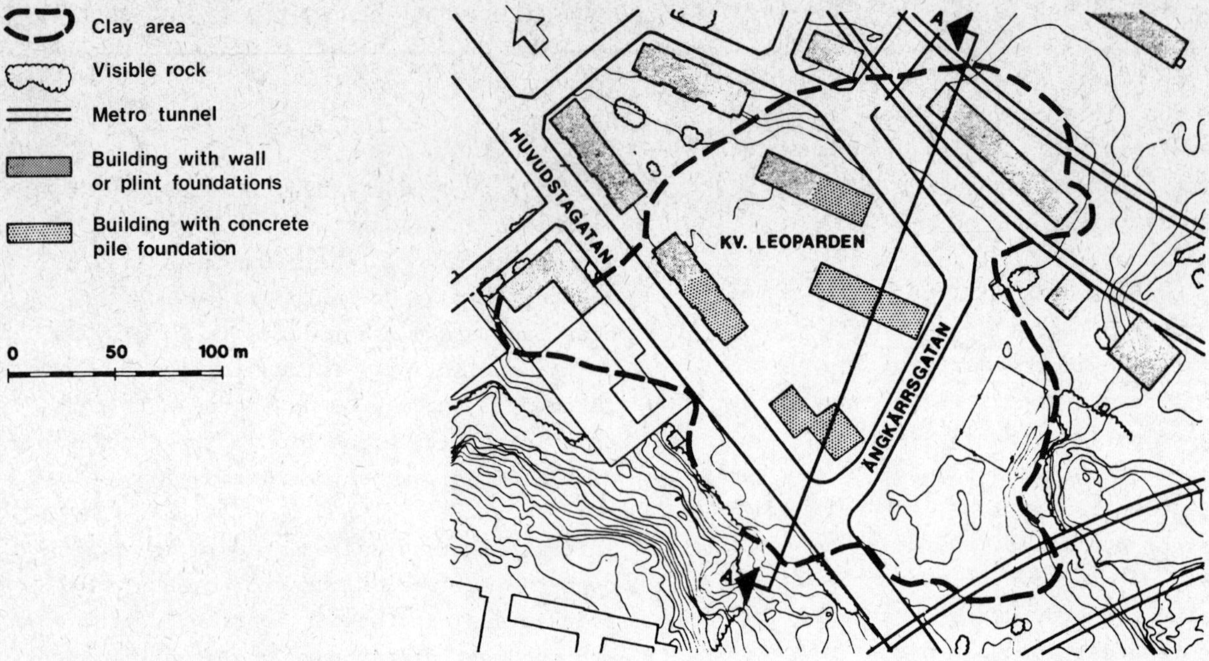

Figure 42

The process of ground water drawdown in a clay depression under a housing area. In the figure can be seen the changes in the level of the ground water resulting from the construction of houses (during the 1950's) and of the metro (1970) and the changes resulting from the preventive measures introduced (artificial infiltration of the ground water). If the preventive measures had not been put inte effect, a ground settling of about 1 m would have occurred over a 25 year period.

Stockholm County Council

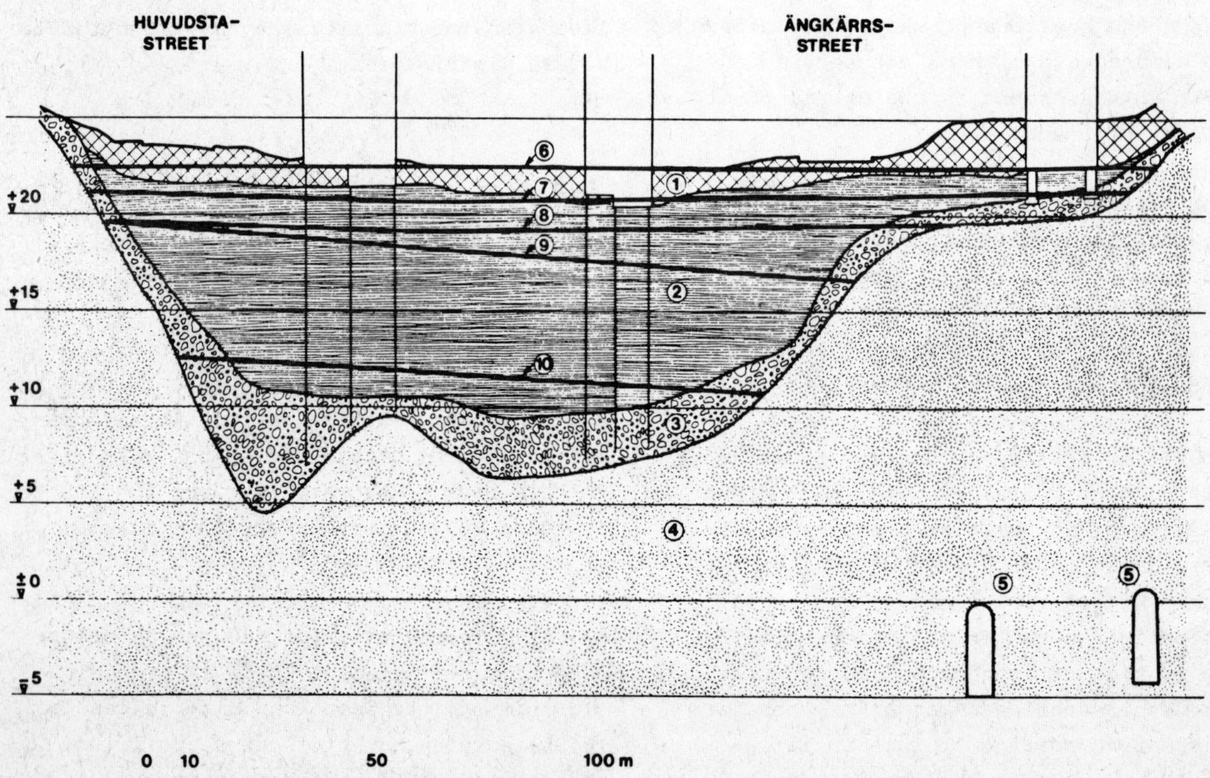

Figure 43

Even if all gravel pits are not vandalised as shown in the picture above, they still constitute a major encroachment on the landscape. Quarries, open mining of rock which is to be crushed, also lead to considerable encroachment.

Planning of the use of surplus material from subsurface installations can provide environmental advantages and at the same time contribute economically to their realisation.

can be sold. This often makes a signficant improvement to the financial aspect of the project. Disposal, on the other hand, is often difficult to arrange, because of environmental effects - permits from the Swedish Environment Protection Board are required - and the cost is considerable. Co-ordination in time between local supply of and demand for earth and rock can therefore be profitable. For example, the metro in Stockholm was built in accordance with a very strict timetable. Because of local demand, the handling of surplus material constituted a large and fluctuating item on the balance sheet.

As an alternative to current extraction of material from gravel ridges and quarries above ground, subsurface construction can make a contribution towards providing a supply of raw material for building purposes. This is excellent from the general environmental standpoint and would also make a financial contribution towards the realisation of subsurface projects. It can be mentioned in parenthesis that the crushing of rock can be carried out in subsurface installations, thus sparing the environment an extremely disturbing activity.

A proposal has been made concerning regional co-ordination between the supply of and demand for rock and earth. This would make possible the rationalisation of simultaneously carried out extraction and dumping of surplus material. The Swedish Environment Protection Board and the Swedish Board of Urban Planning expressed their interest in the proposal in the form of a joint declaration made in 1972. No measure has been published as yet.

9.4 Human beings underground

Subsurface location of certain types of activity implies that human beings spend their time there - for shorter or longer periods. The importance to the individual of spending time underground depends on several different factors.

Subsurface installations themselves are characterised by various limitations in the form of the absence of daylight, reduced accessibility, physical and mental enclosure and also the risk of safety problems, both of a general nature and also dependent on the activities involved, etc. Human beings are influenced to varying extents by environment, this depending partly on the make-up of the person concerned, partly on individual sensitivity. When compared with alternative location above ground level, these factors generally imply a negative evaluation of the subsurface situation. When a choice is involved, however, the advantage factors must also be taken into consideration. In a comparison, subsurface location can imply advantages such as shorter or faster connection systems, protection from rain, wind, cold and other climatic factors, as well as from noisy, dangerous or in other ways undesirable components in the environment, etc. Particular attention should be paid to the good conditions for effective environmental and climatic control.

Justification for the importance of paying attention during the planning stage to the environment in which people are to spend their time can include the following points:

- It is important to avoid poor conditions where people work and spend their time.

  In most cases the conditions are more favourable when these environments are located above ground level. There are, however, many examples showing obvious advantages for the servicing and maintenance of supply systems and processes in subsurface installations.

  From the human being's point of view the motive regarding subsurface installations should be relative to the function concerned.
- Efforts should be made to avoid giving subsurface installations a bad reputation which they do not deserve, with the result that well-motivated utilisation of subsurface installations is hindered.

Decisions made as a result of which large groups of human beings must spend their time in subsurface installations are associated with considerable responsibility on the part of decision-makers. It is therefore reasonable to demand that such decisions are based on a sufficiently wide range of knowledge concerning human reactions to the special subsurface environment and, furthermore, that the subsurface installations concerned are designed in detail on the basis of this knowledge.

This section describes an inventory of the range of knowledge available concerning the physiological and psychological reactions of human beings to spending their time in subsurface installations.

In this survey, extensive work has been carried out in compiling data, not only from different trades and professions where investigations which are applicable in some respect to time spent in underground environments have been carried out, but also from people who have been in contact with the problem. In connection with this work, about 80 interviews have been carrid out, together with a survey of the litterature at about a dozen libraries and data banks.

9.4.1 Current knowledge

Research concerning the risk of injury during the construction stage is today going on at many places in Sweden and also in the rest of the world, throughout commerce and industry, at universities and other training and educational establishments. Since the most serious problems occur during the construction stage, it would appear natural to concentrate research resources to this particular stage.

On the other hand, there only appear to have been a few surveys concerned with psychological and physiological reactions to different components in the subsurface environment during the operational stage. An extensive survey of the litterature shows that knowledge in this respect is poor, even on an international scale. In other words, there has hardly been any research carried out directly aimed at plotting the implications for human beings of spending time and working underground.

General observations

Health surveys and analyses of absenteeism statistics which have been carried out have not indicated any signs that subsurface work of normal duration has a detrimental effect om human beings. This is on condition that climatic conditions with respect to light, noise level, temperature and ventilation have been satisfactory. Several conditions, on the other hand, indicate that time spent underground in itself represents a source of mental stress for many people. Attitude surveys indicate that defects in climatic comfort and social conditions produce a stronger negative reaction in subsurface installations and in installations without windows than in other places. This can be interpreted in such a way that subsurface work in itself constitutes a stress factor, the result being that the tolerance limits for further loading are lowered to an even greater extent. The number of people subjected to such stress loading in connection with time spent underground and the intensity of this loading may, in each individual case, depend on the extent to which those concerned have had the practical possibility of choosing whether they want to spend their time underground or not.

It appears obvious that a selective range of people shoose subsurface work. No surbeys, however, have been carried out as to how many people choose otherwise, or on the factors influencing their choice.

Physiological aspects concerning the subsurface environment

Surveys carried out of workers in subsurface installations have indicated a high frequency of headache, fatigue, eye complaints, nervous complaints and insomnia problems (Forssman Holister 1968). In spite of this no significant differences have been noted in absenteeism due to sickness in comparisons between rock cavity workshops underground and corresponding workshops above ground level.

Simple health surveys have not indicated any differences between workers engaged underground and above ground.

It has been assumed that the complaints concerned have developed from defects, for example, in personnel policy and in the relationship between people in positions of authority and their subordinates at the place of work (Forssman Holister 1968) or from defects in ventilation, lighting, etc. (Trysin).

British, Russian, Australian and German surveys have not produced any signs to indicate that work carried out in conditions without windows causes illnesses. When the hygienic conditions of the environment have been satisfactory, working performance would not appear to have been influenced either. (Holister 1968).

By way of summary,it can be stated that the physiological effect on the human organism of time spent underground has been investigated to a very incomplete extent. Even if no obviously negative effects have hitherto been found, the possibility of the existence of such effects should not be ignored.

Psychological aspects concerning the subsurface environment

The psychological problems to be expected among people who work or spend their time in subsurface premises include claustrophobia, anxiety concerning not being able to get out in the event of a fire, worry concerning roof falls and feelings of discomfort with respect to the knowledge that there are many tons of rock and earth above them. The points most often put forward are varying effects of lack of windows, lack of daylight, similarly of the possibility of airing rooms, and visual contact with the outdoor environment. The experience of human beings in premises without windows has also been the subject of most of the surveys.

It is not possible to state how large a percentage of the population suffers from claustrophobia. Not only have no thorough surveys been carried out, but it is in addition not possible to specify any fixed limits between normal and abnormal reactions. Most people would at least experience a feeling of anxiety if they found themselves stuck between two floors in a crowded lift, but only very few have direct feelings of anxiety when descending to an metro station.

There are various reasons as to why installations are built underground or built without windows: in the centre of cities, the demand for a high level of land utilisation results in premises being built without windows. In connection with certain types of manufacture and research, it is important to be able to control the noise level, the lighting, the temperature or the quality of the air. In facilities such as museums, theatres and schools, the intention can be to eliminate sources of distraction.

The activities carried out in these types of installation vary within wide limits and it is by no means unexpected to find different attitudes to the lack of windows depending on the type of activity studied (Holister 1968, Collins).

It appears that surveys have been carried out which aim to determine the way in which reactions to lack of windows are influenced by such factors as the time spent in such premises and the type of employment concerned, but it is still quite clear that such influence exists.

Studies of human beings in different types of premises without windows indicate clearly that the highest level of dissatisfaction occurs in small premises with limited activities and with limited opportunity for the establishment of contact between individuals. Examples of this include small rooms in nursing wards where individuals are restricted to their beds and have nothing to do. Routine jobs in offices and industry, where the employees enjoy a limited range of movement and where the activities are carried out within limited space, also belong to this category.

Department stores and metro stations, on the other hand, are characterised by large areas with a high level of activity and great variety, and involving many people. In dynamic environments such as these, the lack of windows and daylight does not give rise to any obvious expression of dissatisfaction.

The conclusion to be drawn from studies carried out up to now is that the following factors make up the most important reasons for expression of dissatisfaction in connection with time spent in a subsurface environment:

- boring and monotonous jobs
- limited range of freedom
- limited social contact
- confined areas

These factors constitute a static and limited environment which creates a certain level of stress, this apparently being further underlined by the absence of daylight.

Furthermore, it would appear as if varying and stimulating jobs can, to a certain extent, counteract various defects in the physical environment.

9.4.2 Characteristic factors

Without any direct support being available from research or other sources, listed below are different factors which are of prime importance where subsurface installations in which time is spent are concerned:

Subsurface installations are generally characterised by:

- lack of windows
- reduced accessibility
- constant climatic conditions

Figure 44

Artistic decoration of metro station in Stockholm

Svenska Vägaktiebolaget

Further environmental factors influencing the experience of subsurface time:

- the practical possibility of choosing the subsurface environment
- the social status implied by subsurface activities
- the length and frequency of occasions when time is spent there. In the case of time being spent there permanently, the possibility of taking rest periods above ground level.

The design and utilisation of subsurface installations:

- the distance to the surface of the ground measured in time and distance. The experience of distance is important.
- the planning of connecting passages. Horizontal or sloping entries can give the impression of easy accessibility, more dependable evacuation facilities being implied as compared with stair wells and lifts.
- the choice of dimensions and the design of rooms with respect to
  . variation in reaction, sense of orientation
  . spaciousness, possibility of surveying entire premises
  . the possibility of establishing social contact during the operational stage.

Equipment in subsurface installations

- the design of the entry section and the connecting passages. A soft and well-furnished transition from ground level to subsurface installations is important from the point of view of the people experiencing transition
- variations in room design, planned colour compositions, artistic design of floors, walls, and ceilings and also interior fitting units
- the acoustic characteristics of the room. Uncovered rock installations are difficult in this respect
- lighting conditions
- climatic control. Subsurface installations are very easy to handle in this respect.

9.4.3 The situation today

As has already been indicated, no detailed knowledge is available concerning the reaction of human beings to the subsurface environment. Furthermore, there would appear to be a lack of compiled information on how many people on the whole spend their time in subsurface installations in Sweden, as well as on the extent to which and the conditions under which this occurs. The number of people active in subsurface installations, excluding the mining industry, can be roughly estimated at between 5,000 and 10,000 people. Current types of work place include the following:

- industrial installations
- defence installations
- power production installations
- the metro in Stockholm
- department stores
- technical service supply systems

The number of people making occasional visits to subsurface installations in connection with travelling on the metro with commercial and service facilities, etc. is considerably greater.

Furthermore, there are no standards to aim at concerning the special subsurface conditions. On the contrary, difficulty in defining cavities as buildings from a legal point of view prevents the exercising of authority (see page).

Laws have been created, on the other hand, to give employees influence concerning layout of the environment. The Industrial Safety Laws, Industrial Safety Bulletins and Building Regulations which came into effect in Sweden on January 1, 1974, specify, for example, that industrial safety officers, industrial safety committees or orginisations representing the employees have the right to participate in the planning of premises. In order to be able to grant building permits for working premises or staff facilities, the Swedish Factory Inspectorate must be provided with a statement indicating that the employees have had the opportunity to express their opinion on drawings and other documents.

In cases where it is not known who is to work in the premises concerned, during the planning and design stages the Swedish Factory Inspectorate and the regional industrial safety representatives must be afforded the opportunity to examine the project and to make decisions as to whether the subsurface location is advisable from the environmental point of view.

One reason why the subsurface environment has attracted so little interest may be that employees working below ground level are divided up to a great extent in many small places of work or into small units in larger organisations from which these units are separated. This may imply difficulty for employees in subsurface work to put forward their special demands and viewpoints concerning the working environment through their trade union or industrial safety organisation.

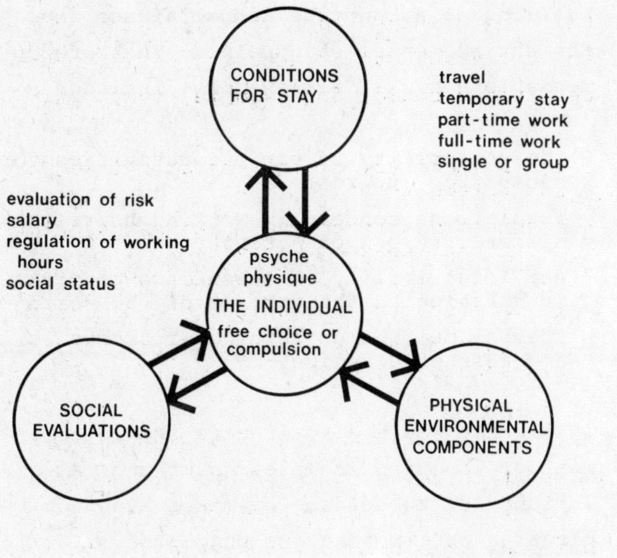

Figure 45

General reference outlines for evaluation by the individual of time spent in subsurface installations.

9.4.4 General reference outlines for analysis of the subsurface environment

The general reference outlines shown in Fig 1 have been drawn up as a basis not only for analysis of the influence of the subsurface environment on the individual but also for evaluation by the individual concerned of the subsurface environment itself. The reference outline also provides a basis for discussion on the insufficiency of knowledge available and on the need for research. The reference outlines illustrate - in a very simplified way - the interplay between the individual and the functional, social and physical environmental factors surrounding him or her. Important or characteristic conditions for time spent underground can be obtained from each main group of environmental factors.

9.4.5 A suggestion concerning the formulation of research requirements

It is suggested that future research work within this range of problems be divided into the following three phases:

PHASE 1 Plotting of the extent of subsurface work and environmental problems

PHASE 2 The effects of the subsurface environment on the human being from the physiological and social angles

PHASE 3 Formulation of planning recommendations and the development of technical solutions

PHASE 1

The aim of this plotting would be to provide a review of the different types of subsurface installation and subsurface work. This must cover not only actual work places, but also places that people occasionally visit and spend time in below ground level. A general plotting of this type could provide the basis for choice of installations and subsurface work for closer study.

The way in which employees and temporary visitors experience the subsurface environment and the problems pertaining to it could be plotted with the help of such features as interview surveys. A study could also be made of the way in which the problems are related to the type of work, the time spent below ground level, the layout of the installation, etc. If data were collected from comparable installations above ground level, differentiation could be made between general environmental problems and environmental problems dependent on location underground.

The compilation of data in this phase could cover the following:

a. Information concerned with the occurrence of subsurface work, etc.
   - the number of people employed in subsurface work
   - division as between trade and profession and also time spent in the subsurface
   - the occurrence of people working alone or of work implying special health risks
   - the number of people occasionally paying visits to subsurface installations

b. Environmental conditions in existing subsurface installations with respect to
   - space, area
   - accessibility
   - noise
   - light
   - air
   - safety
   - aesthetic and architectural standards

PHASE 2

The occurrence of stress symptoms and psychosomatic problems in various types of subsurface work could be investigated, comparisons being made with similar working situations above ground level. For example, absenteeism due to sickness and staff turnover for various trades and professions underground could be used as a gauge of mental stress involved in the work.

A study could be made of the implied significance of the specific environmental capacities of subsurface installations with respect to the way in which people experience mental stress. These characteristics could then be related to the type of work, the time spent there, possibility for social contact and the physical stress involved in the work.

A study concerning shorter periods spent in subsurface installations could generally be accommodated in a more extensive study of the working environment.

PHASE 3

The third phase would cover the formulation of recommendations for the planning and building of subsurface installations from the environmental standpoint. These recommendations could, for example, include:

- the suitability of various activities for subsurface location
- regulations concerning working hours in different types of activity
- the accessibility of subsurface premises in relation to the surface of the earth
- special demands concerning architectural and aesthetic design

This phase would also involve the development of layout principles and technical solutions for subsurface premises based on the planning recommendations suggested.

9.5 Construction costs

When a subsurface alternative is to be considered various skills are required as regards construction techniques. Examples are as follows:

- How can subsurface accommodation be designed under various geological and hydrogeological conditions?
- What demands will be made on implementation, for example, to avoid damaging buildings in the vicinity?
- How long will the construction work take to complete?
- What costs are associated with the project?

A fairly good idea of an individual project can be obtained primarily on the basis of thorough preliminary surveys. Even when these have been carried out, there is normally still a certain degree of uncertainty since it is not possible to obtain comprehensive knowledge about the structure of the ground for reason of cost. For this reason subsurface construction implies a certain amount of risk since the construction work can be more complicated than anticipated due to unexpected local anomalies in the rock.

It is possible, on the other hand, to make a general estimate of the technical solutions

Figure 46

Atlas Copco

Figure 47

Skanska

Figure 48

Stabilator AB

that are possible and also the associated costs. This type of estimate is of the greatest value in a planning situation.

Only the cost aspect is considered in this section. A proposal is submitted as to how construction cost for recurrent types of work can be calculated on a general basis.

9.5.1 Conventional method (drilling/blasting)

General costs

The cost of excavating tunnels and rock cavities depends on many factors such as:

- the quality of the rock
- the layout of the installation
- the extent of the work (tunnel lengths, cross section areas, etc.)
- the construction period
- the depth below the surface of the ground
- available parking for machines, equipment etc.
- the transport distance and the number of supplements
- possibilities for tipping and disposal of surplus excavated material
- possible restrictions concerning blasting with respect to buildings, etc.

The actual excavation costs can be calculated relatively easily, but it is more difficult, on the other hand, to be able to anticipate the required amount of reinforcement and thereby the associated cost. Sometimes they can be excluded all together, while in other cases they may make up the major part of the cost in connection with tunnelling. (This makes it important in design work to have basic material from preliminary surveys which is as comprehensive as possible, so that reinforcement work can be anticipated with a satisfactory degree of probability. Only then can a firmer grasp be obtained concerning the building cost aspect in connection with subsurface work.

Reinforcement work can interrupt and disturb the normal cycle of work and time can be lost. Sometimes it can be such interruptions, and not the reinforcements in themselves, which are decisive from the cost aspect. The following account includes only the direct costs for reinforcement work.

9.5.2 Model for general calculation of installation costs in connection with subsurface construction

Below is shown a list of the various subsidiary costs in connection with construction work and also a general layout for compiling the total cost. The information on costs has been derived from various consultants and contracting companies and also from the city engineering authorities in Stockholm and Gothenburg. The costs refer to the 1973 level.

Tunnelling costs

Tunnels

The excavation costs for tunnels in the size range of 5-100 m² using conventional methods is indicated in Fig. 51. The cost is based on a theoretical solid volume and includes - apart from the actual excavation costs - loading and transport out of the tunnel of the excavated material. In addition, there are the following costs:

- design work, administration, control work and contingencies	25-35%
- reinforcement and sealing, normal (can vary to a considerable extent)	10-20%
- transport tunnels, raises, sunk shafts	-
- bulk transport above ground level	-
- rock cavity index, annual increase rate at present	approx 15%

Rock cavities

The cost of excavating rock cavities with cross-section areas greater than 200 m² is about Sw.Crs 40/m³. To this must be added extra costs as in the case of tunnels. The calculation model is shown in Fig. 52.

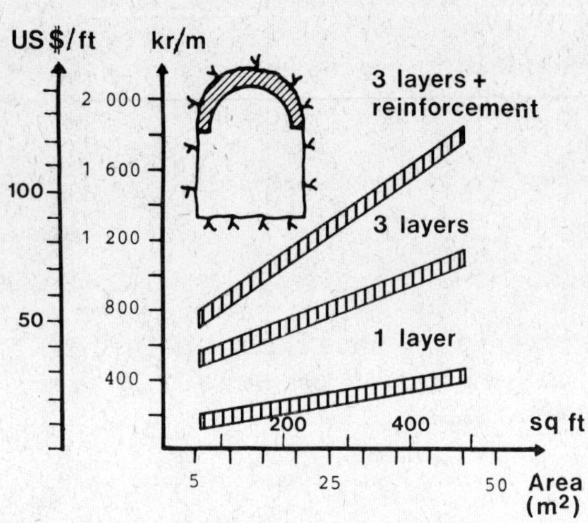

Cost for shot-creting ceiling as function of tunnel width.

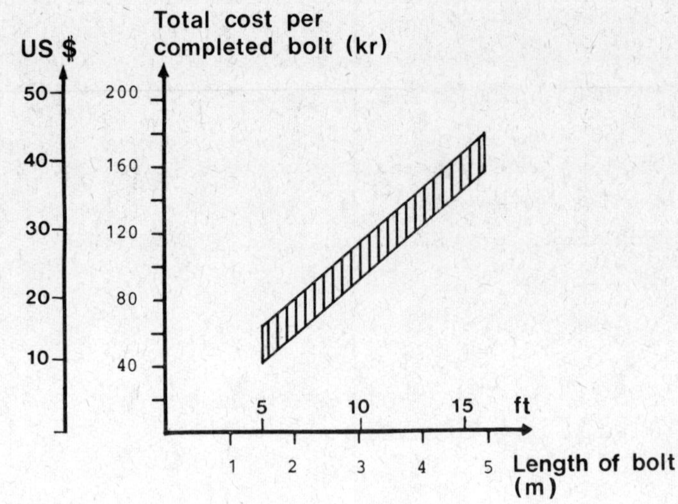

Cost for completed bolt as function of bolt length.

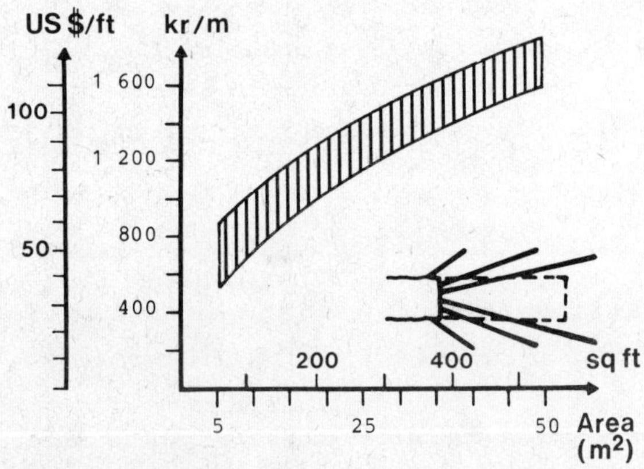

Cost for pre-grouting as function of tunnel area.

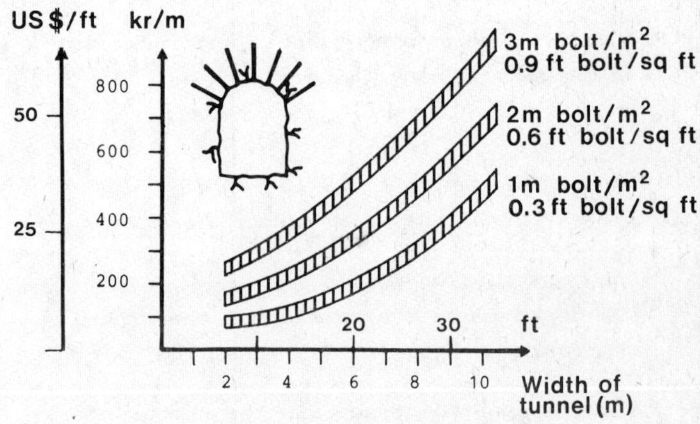

Cost for systematic bolting as function of tunnel width.

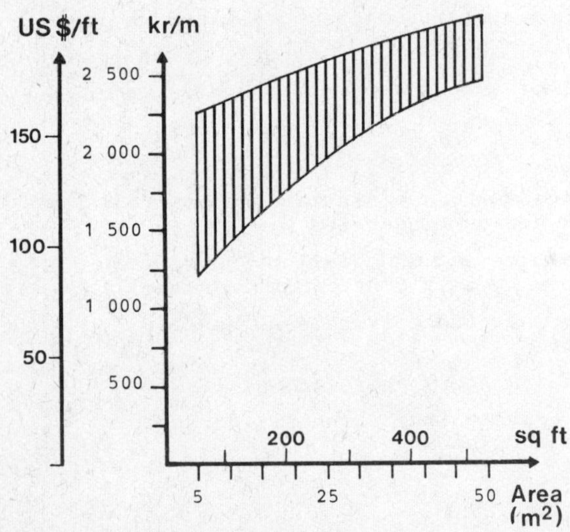

Cost for post-grouting as function of tunnel area.

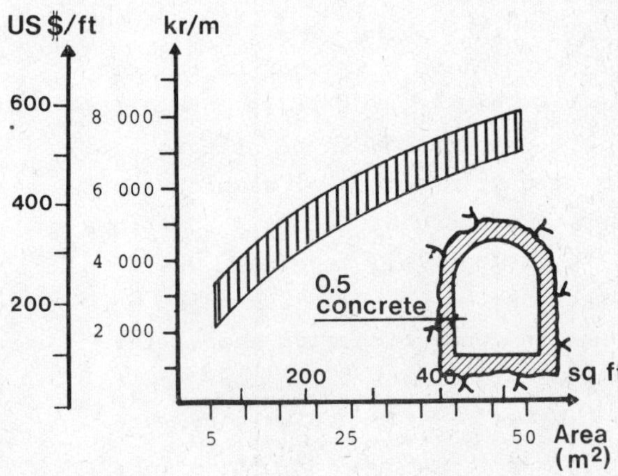

Cost for lining as function of tunnel area.

Figure 49

Costs for various reinforcement and sealing work in US$ and Sw Crs. Price level 1973.

Vattenbyggnadsbyrån

Raises - Sunk shafts

Excavation costs are about 2 - 3 times and 3 - 4 times respectively the cost of excavating a horizontal tunnel with the same cross-section area.

Reinforcing and sealing costs

The costs for various reinforcement and sealing work are described on the opposite page. The graphs do not meet any great demand on accuracy, but merely show the approximate magnitude of the costs for various operations.

Calculation model

The total construction costs can be estimated by using the calculation models in fig 51 (for tunnels) or in fig 52 (for rock cavities). Note that communicating tunnels are allocated higher prices since they are not horizontal.

There has recently been a marked increase in cost development, see fig 50 At the same time continuous improvements are being applied within the field of tunnelling and reinforcement techniques. Other factors involved include the demand made on accomodation, which varies according to the intended range of use (for example, if human beings are to spend their time there) with the risk of environmental consequences, etc.

If such cost information is to be of practical value, there must be continuous follow-up of the technique and cost. (When the total construction cost for an alternative project is to be calculated, the calculation must also include measures to be carried out above ground level, for example foundation reinforcement, etc.).

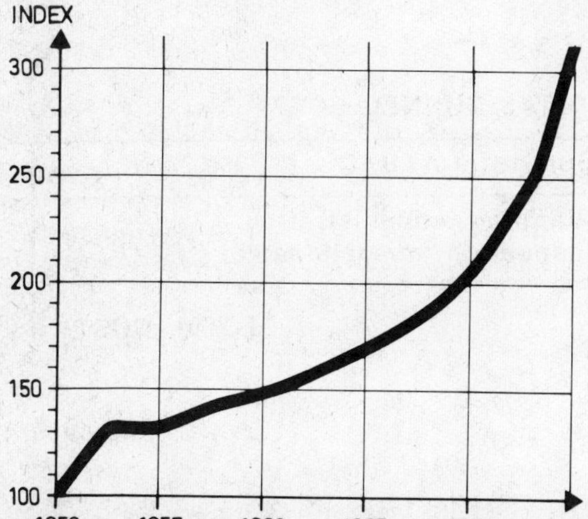

Figure 50

Development of tunnelling costs

9.5.3 Other tunnelling methods

Since these methods are not so usual and since special conditions prevail each time they are used, it is difficult to specify costs in absolute figures. An impression of the costs can, however, be obtained by specifying the ratio between different methods for tunnels with the same cross-section area. The basis is a rock tunnel built in a conventional way which gives a unit cost of one. Here are some examples of relative costs for various types of tunnel.

Conventional tunnelling	1
Full-face drilling	3
The "shield" method	10-15
Cut and cover	3-4
Sunk tunnel	10-15

9.5.4 Needs for reforms

As was mentioned in the introduction, knowledge of the conditions as regards construction technique is important so that the possibility of siting activities in the subsurface can be given the attention it deserves in the planning process. This means that knowledge that can be handled as outlined above is valuable. The need for knowledge, however, also applies to other characteristics than costs. In other words, informa-

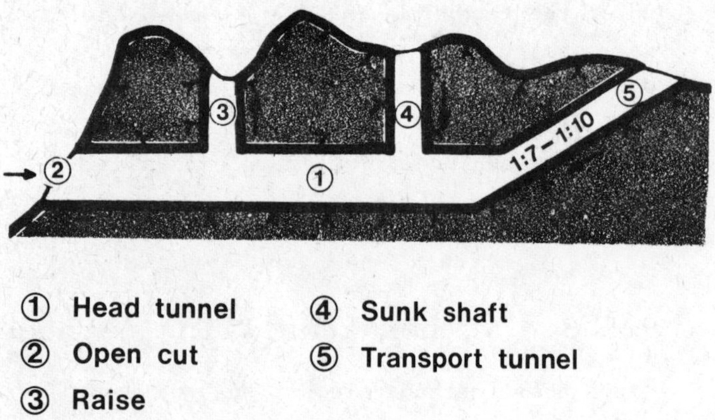

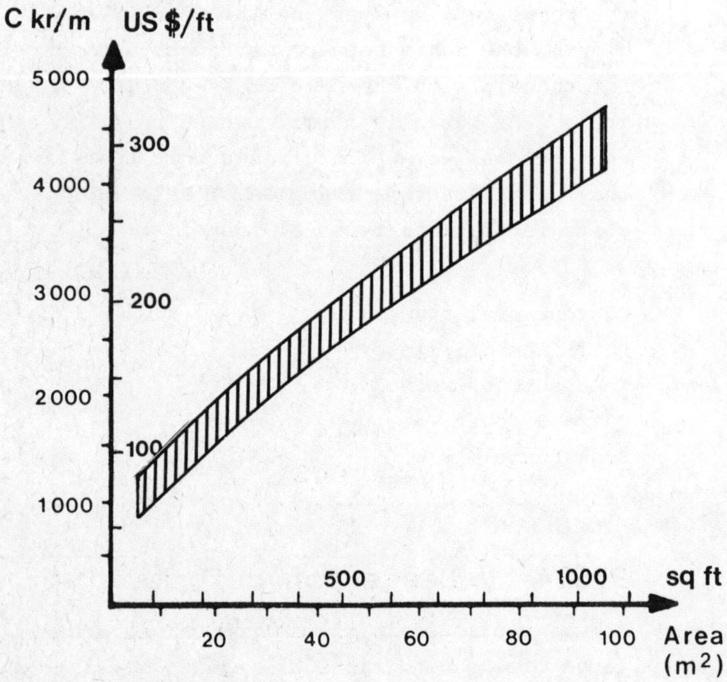

A. Costs of rock blasting
(K= price per unit acc. to fig.)

Total A

B. Costs of handling of blasted rock

Total B

C. Costs of strengthening and sealing

Total C

TOTAL TUNNEL COST

Sum total A+B+C	
Planning, administration Inspection, miscellaneous	+25–35 %

TOTAL COST

Figure 51

Calculation model for tunnel costs

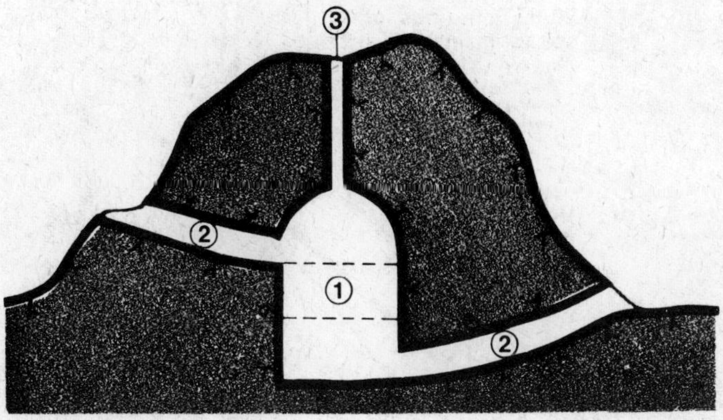

① Rock chamber

② Adits

③ Raise

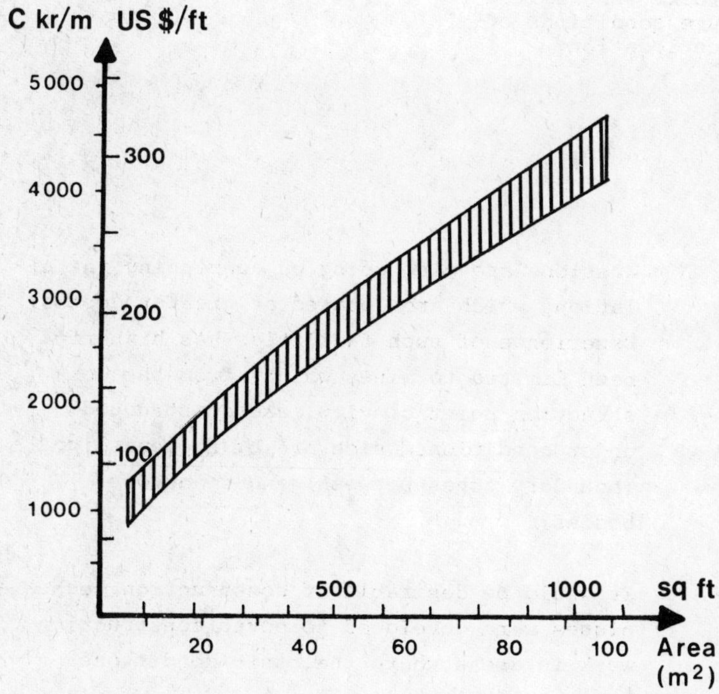

A. Costs of rock blasting

Total A ☐

B. Costs of handling of blasted rock

Total B ☐

C Costs of strengthening and sealing

Total C ☐

TOTAL TUNNEL COST

Sum total A+B+C		
Planning, administration Inspection, miscellaneous	+25–35 %	
	TOTAL COST	

Figure 52

Calculation model for rock cavity costs

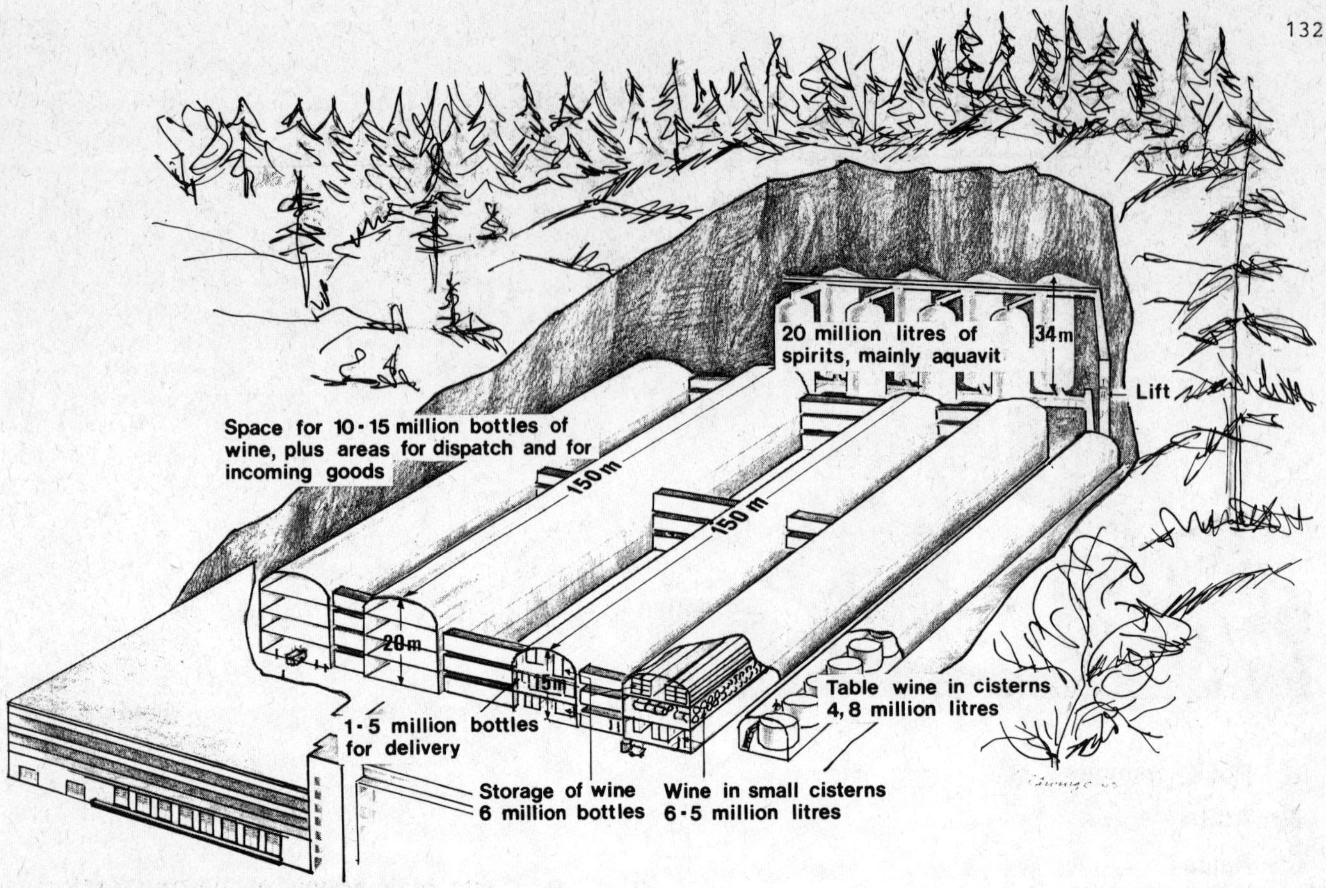

Figure 53

The storage areas at Vin-& Spritcentralens installations at Årstadal in Stockholm are mainly located in rock. The limited availibility of land and the steady temperature conditions of the bedrock were the reasons behind subsurface location

tion should be produced and kept up-to-date for all relevant aspects. The statistical follow-up suggested in section 9.1.7 will be of great value in this respect.

As far as practical application is concerned, research work is today proceeding in several different directions. Studies for the subsurface location of nuclear power stations, for example, have indicated that it is quite possible to carry out a span of 50 m under favourable conditions. Great interest is being devoted to installations with extremes of loading, pressure or temperature. Discussions are also going on concerning installations which are located at greater depths. Experience of such facilities has hitherto been limited to mines which, from the construction point of view, are constructed under conditions which are by no means ideal (boundary zones between gangue and ore bodies).

It would be desirable if construction techniques were developed to cover construction work in areas where the basic conditions are not ideal.

9.6 Cost-benefit appraisal

9.6.1 The concept of cost-benefit

The concept of cost - benefit in appraisals, analysis and evaluation refers to the joint effect on the project and on decisions concerning the project of factors which are measurable, difficult to measure and unmeasurable. Using this wide interpretation of the concept,each stage in the development of a plan or project implies some form of cost - benefit appraisal. One basic characteristic of each appraisal of this kind is an attempt to apply systematic attention to all the important factors and to carry out a systematic compilation of available facts and experience as well as opinions.

Cost - benefit appraisals are characterised by extreme complexity. Among other things, the different parties concerned act with different interests and objectives, which often leads to differences in appraisal. In general, three parties or groups of parties can be differentiated, as follows:

The company	- Responsible for the project or functions
The society	- With overall responsibility from the point of view of the general public
Interest groups	- Depending on the geographical location and the type of project, these can consist of environmental groups, political or corresponding idealistic organisations, land-owners, etc.

One characteristic of planning activities today is that cost - benefit appraisals from the point of view of companies in connection with society cost - benefit appraisals are often dominated by social and environmental evaluations and that the appraisals of the interest groups have a marked influence of the overall appraisal, supported by the development of opinion and by the mass media.

The administrative treatment of plans and projects is not adapted to suit the balance of power at present prevailing between the parties concerned. Credibility gaps easily occur between the three parties consisting of companies - society (the establishment) - interest groups.

The situation can result in plans and projects which have been developed according to current administrative routines being revised, blocked or stopped at the stage when the design work has been completed or construction work has even been started. The result is a disastrous waste of resources in the form of both labour and materials.

One recommended way out of the present planning crisis is to give consideration to principles at an earlier stage, to make long-term planning more important and to formulate objectives and joint planning conditions in which the interests of the various parties are thoroughly taken into account and balanced mutually.

Consequently, cost - benefit appraisal should already be carried out in connection with overall planning and partial planning with due to attention given to the interests of the various parties.

Demands are made on the technique of cost - benefit appraisal so that generalisations and simplifications are used to produce easily handled basic material for decisions. Particularly at an early stage of planning, it is therefore important to have access to general information on the different fields of technology and on the collected experience as to its use.

This research project has shown that there is a great lack concerning this type of information with respect to the construction and the utilisation of subsurface installations. Particularly in connection with cost - benefit appraisal, but also in connection with rational utilisation of the subsurface in general, there appears to be a great need for compiled information.

Some institution, probably state-owned, should develop interest in or be obliged to compile and process such information.

Many analysis methods have been developed and used for cost - benefit appraisals, particularly from the point of view of companies with an emphasis on purely financial evaluation. Examples of methods which can be used for planning and for the determination of results are the following:

. Analysis of functional requirements (systematic plotting of such features as technical aspects).

. Analysis of energy costs

. Analysis of transport work

. Check-list of criteria, identification of aspects and descriptions in a systematic way.

. Utility analysis, that is to say, degree of goal attainment.

. Resource balance analysis (the balance between various material and financial resources processed through real-economic and financial investment calculations.)

. Quantification of the consequences in real terms, i. e. expressed in other ways than in money (for example in man-hours, the number of different people affected).

. Sensitivity analysis (variation in result if some parameter which can or cannot be influenced is assumed to be changed.)

Different methods have their advantages and disadvantages. It is essential that cost - benefit appraisals be as comprehensive as possible with respect to influential factors and that the analysis work concerned, from the point of view of extent and degree of detail, be in reasonable proportion to the value of the expected result.

9.6.2 Comments in connection with company appraisal

Appraisal of future development of a particular function is formulated in the form of objectives. The objectives specify not only the minimum requirements but also the varying degree of desired or anticipated goal achievement. Systems or projects, the realisation of which means that minimum demands are satisfied, make up an alternative for comparison analysis. This situation for appraisal is characteristic when the choice concerns surface or subsurface alternatives or a choice between different subsurface alternatives.

Many types of comparison analysis have been studied within the framework of this research project. They are of a typical business economics nature and the intention has been to evaluate short- and long-term costs as compared with financial benefit for the various alternatives.

Four groups of projects have been studied, as follows:

. Pipelines and cables

. Traffic arteries

. Industrial installations

. Stores

Four examples of the compilation of financial calculations are described here:

1.

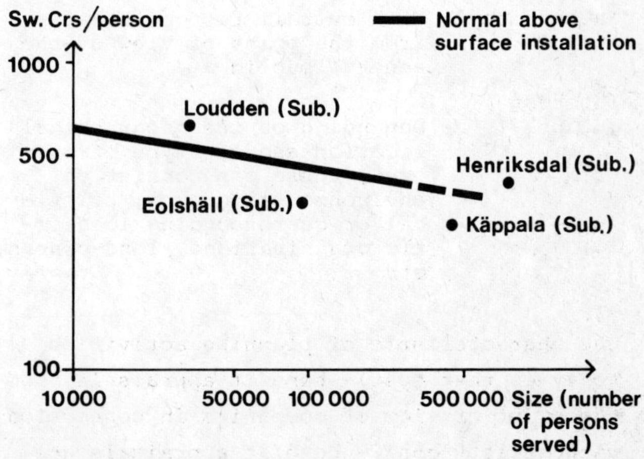

Figure 54

Investment in sewage treatment plants of various sizes, above surface and subsurface. 1970 price levels.

2.

	Subsurface garage in rock	Traditional above surface garage	Above surface lorry park
Investment, total excl. heating	8.5 Sw Crs/M	8.5 Sw Crs/M	5.5 Sw Crs/M
Investment in land (included in above)	0	2.2 "	2.2 "
Investment per vehicle bay	40 000 Sw Crs	40 000 Sw Crs	25 000 Sw Crs
Rent per vehicle bay and year (excl. heating, water, administration)	4 000 "	3 600 "	2 400 "
Rent per m^2 and year	54 "	80-103 "	66 "

Comparative calculations have been made for a garage in a rock chamber under a park in central Stockholm and for an alternative location above ground at a suburban site. The installation is for 200 lorries and include a petrol and repair workshop with space for the vehicles.

4.

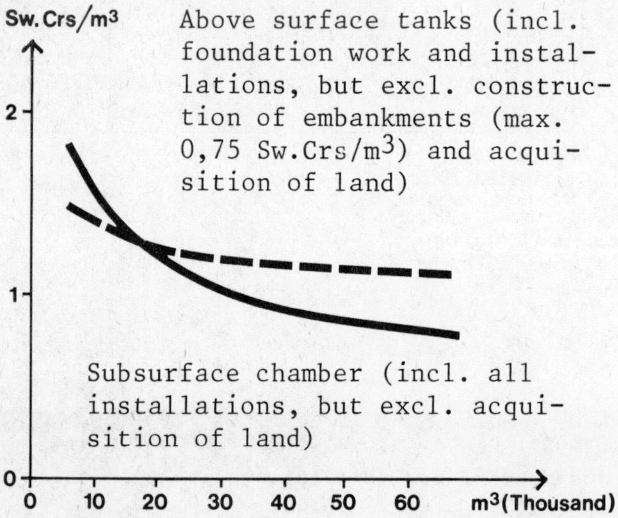

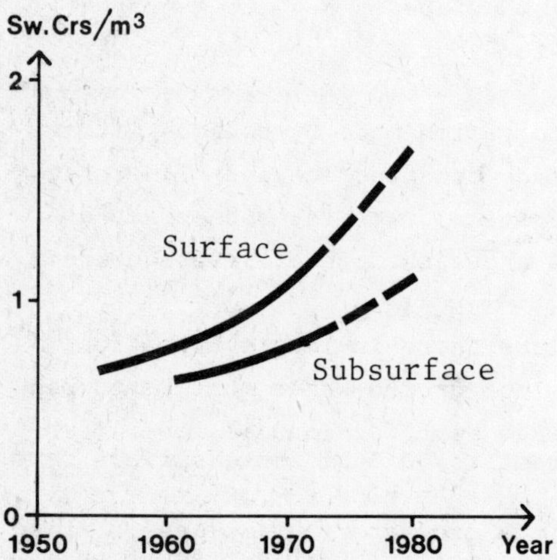

The profitability limit between above surface and subsurface siting of oil storage was in the 1960's at approx. 40,000 m³. Today it has come down to a storage volume of 20,000 to 30,000 m³. The relationship between investment and storage volume is as shown in Figure A. Development in time of the cost differential is as shown in Figure B.

3.

	Tunnels		Surface conduits	
	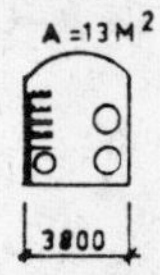 A = 13 M² 3800	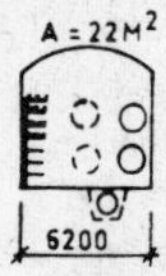 A = 22 M² 6200	Newly built areas	Town centre
	Sw Crs/M	Sw Crs/M	Sw Crs/M	Sw Crs/M
District heating	2 500	3 300	2 300	3 400
Water	900	1 000	600 1)	900
Cable	1 500	1 500	600 1)	1 500
Total	4 900	5 800	3 500	5 800

1) Costs are based on the laying of pipe and cable respectively in conjunction with road construction

Stockholm 1975. Costs for conduits at surface level and in multi-use tunnels

Tunnel installation			Prel.calculation		Cost account		Appr. quotient for the VAST[x)] index for rock work	(6)/(4)	Comment on diff. between prel. calcul. and cost account in fixed value of money	(9)/(8)
No.	Area m^2	Length km	Sw.Crs mill.	Year	Sw.Crs mill. current value of money	Year of constr.				
(1)	(2)	(3)	(4)	(5)	(6)	(7)	(8)	(9)	(10)	(11)
1.	210	0,3	1,3	1950	1,2	1951-5	1,2	0,92	Considerably cheaper than estimated	0,77
2.	260	3,4	20,6	1954	23,1	1953-7	1	1,12	Somewhat more expensive than estimated	1,12
3.	260	0,6	4,1	1955	3,9	1957-9	1,02	0,95	Somewhat cheaper than estimated	0,93
4.	30	5,7	5,9	1966	16,1	1966-8	1,03	1,03	As estimated	1,00
5.	45	0,45	0,6	1966	10,6	1968-9	1,11	1,00	Somewhat cheaper than estimated	0,9
6.	65	4,4	8,0	1971	11,2	1971-3	1,1	1,40	a)	1,3 a)
7.	45	3,84	6,20	1971	6,78	1971-3	1,1	1,09	b)	1,0 despite b)
8.	33	5,4	7,0	1972	8,0	1972-4	1,1	1,14		1,00

a) Costs for reinforcement accounted for 44 % instead of the estimated 15 %
b) Costs for reinforcement accounted for 21 % instead of the estimated 15 %

The above comparison has been obtained for a number of Swedish tunnels. On average, the preliminary calculations and cost accounts for these installations were the same if calculation is based on fixed value of money with the aid of the VAST[x)] index for rock work.

x) Swedish Power Association, Development Section

Cost - benefit appraisals have been carried out on the basis of preliminary calculations of investments, operating costs and construction time. A certain amount of basic material has been produced to relate preliminary calculations to cost accounts. In general, where subsurface installations are concerned, the cost difference considered at a fixed value of money, can mainly be put down to difficulties in initial appraisal the quality of the rock and probable ground-water flows.

It is often difficult to obtain information on cost accounts for two reasons, as far as can be seen. Such accounts are not always made up and those concerned appear to be ashamed that the preliminary calculation was incorrect.

One example of comparison between preliminary calculations and cost accounts with reference to investments is described here in connection with eight tunnel projects.

General operating experience from surface and subsurface installations has been collected here, primarily from oil storage facilities. In the following list, the operating costs for subsurface stores are compared with those for stores above ground level:

According to claims made by sources within the subsurface construction and oil storage fields, operating costs for subsurface as compared to above surface oil storage are:

- lower, since there is less evaporation
- higher, since ground-water must be pumped
- lower, since rent for land is lower (subsurface rent is 50 % of above surface rent per m^2 of surface area and year)
- lower, since products require less heating (subsurface cost is 50-60 % that above surface)
- lower, since maintenance cost for installations is lower (subsurface cost is 0.25 % that above surface)
- lower, since insurance premiums for installations are lower (subsurface premiums are 7-10 % those above surface)

- lower, since insurance premiums for products are lower (subsurface premiums as a percentage of those above surface: for petrol, 35 %; for heavy oil, 45 %; for crude oil 30 %.
- lower, since fire safety precautions are simpler
- higher, since there is increased pumping of oil

Furthermore, risk in connection with earthquakes is less subsurface than above surface. The same applies to the risk that empty tanks might blow away in high winds.

In other respects, the material studied does not permit any form of general evaluation. Individual examples indicate lower costs for the maintenance of pipelines and cables in tunnels, relatively large savings in running costs in the case of subsurface cold stores, on condition that the installation is designed to meet subsurface conditions, and also higher operating costs for subsurface road tunnels.

Experience with respect to the time needed to complete the project should be given due attention in cost - benefit evaluation. The construction time needed for a subsurface installation, which often is longer than that for a plant above ground level, means, on the one hand, higher interest costs. On the other hand, the longer construction period can be more than compensated for by extremely simplified and time-saving preparation of the project without any conflict with existing or planned land utilisation.

For example, subsurface oil storage installations generally require a shorter construction time than corresponding facilities above ground level.

Due attention should also be paid to differences in the conditions for financing through loans with real estate mortgages as security. The granting of loans for subsurface installations occurs more seldom and therefore no system of rules has been developed.

The research project has not traced any complete cost - benefit analysis with respect to subsurface construction in which an attempt has been made to embody social, environmental and communal benefit aspects in addition to business economics evaluations. However, it should be in the interest of those responsible for particular functions to provide as comprehensive and as just a picture as possible in cost - benefit appraisal. The conditions on which those factors not associated with the companies concerned are to be included are that the community plays an active part in the formulation of objectives and also in long- and short-term planning, which also includes the use of the subsurface.

Cost - benefit analysis really refers to a more or less absolute evaluation as to how the sacrifice of resources is in a reasonable proportion to the performance level attained by the final installation.

The conditions for the achievement of high quality cost - benefit analysis include the following:

- Precise formulation of objectives for particular functions and of the implications of these.
- Dependability in the description of the project or the system and its consequences.
- A suitable model for analysis which has been adapted to the prevailing degree of detail in planning, that is to say detail planning, partial planning or overall planning.

9.6.3 Community appraisal

Cost - benefit appraisal by the community of a subsurface installation or the utilisation of a subsurface installation should include a number of factors which are outside the area of responsibility of the particular function. In connection with appraisal, there is also reason to vary the amount of importance attached to certain factors com-

Figure 55 ANNUAL ECONOMIC BENEFITS ACCRUING FROM TRANSFERRING CIVIL WORKS FUNCTIONS TO SUB-SURFACE SPACE IN BILLIONS OF DOLLARS

Cost Category \ System	Shelter	Production Systems and Industrial Structures	Water Resources	Waste Water Control	Solid Waste	Transportation	Communications	Energy Distribution	Total
Direct Costs	2.8	3.5	1.0	1.0	2.0	2.3	0.5	1.2	14.3
Time	0	4.0	0	0	0	8.0	0	0	12.0
Land	1	0.3	0	0	0	1	0.1	0.1	2.5
Energy	3.8	3.8	0	0	0.1	0	0	0	7.7
Pollution	0	0	0	1.3	0.1	4	0	0	5.4
Safety	3	0	0	0	0	3.1	0	0.1	6.2
Reliability	0	0	0.1	0	0	0.2	0.1	1.0	1.4
Material Resources	3	2.5	0.8	0	0.8	1.0	0.5	0.5	9.1
Total	13.6	14.1	1.9	2.3	3.0	19.6	1.2	2.9	58.6

pared with the case in business economics appraisal.

Apart from business economics evaluation, attention should be paid to the following:

- effects on employment
- benefit to public finances, such as income from taxation or reduced public expenditure.
- the effects on the natural environment.
- the effects on land and buildings in the vicinity.
- the effects on land utilisation, the maintaining of existing utilisation and a range of freedom for utilisation in the future.
- the effects in connection with work below ground or other reasons for time spent below ground.
- safety or protection against disturbance from the inside or from the outside.

For guidance in the formulation of a general opinion and in concrete planning situations, a general form of applicability appraisal would be valuable with respect to which activities can be located below ground with greater of lesser benefit - depending on the case in point.

The American survey "The use of underground space to achieve national goals" is an example of an appraisal of this type. The material in the survey, an example of which is shown in the figure, is particularly comprehensive. The basis for appraisal was that the costs and time for subsurface construction could achieve parity with the costs and time for construction above ground level, as a result of concentrated research and development efforts.

The level of ambition in such an applicability appraisal can vary depending on the number of factors included, the degree of detail and variations in conditions.

The advanced applicability appraisal mostly concerns complex systems, that is to say, where all the elements included in alternative systems are described, calculated and evaluated in a versatile way. The simple case can be illustrated with the question as to whether to locate oil storage above or below ground level. The complex system can be illustrated by a complete utility supply system for an urban area.

10. WHAT ADMINISTRATIVE ROUTINES ARE REQUIRED?

10.1 The right to establish subsurface installations

10.1.1 Requirement for planning

The first section of the Building Law contains regulations which state that the use of land for building shall be preceded by planning and that permission (building permit) is required for building. The law is written with building above surface in mind. There is no requirement that subsurface installations be preceded by the careful planning which the Building Law otherwise stipulates. Since subsurface installations, as previously stated, often require very extensive, construction work with special risks as regards the environment, it would seem natural to draw up regulations for the integration of subsurface planning into other community planning.

The planning of subsurface projects and utilisation of the subsurface can, however, not be improved unless the right and obligation to plan subsurface projects and utilisation of the subsurface is written into the Building Law. Only in this way will there be the real preconditions for co-ordination of objectives, needs, use of resources and priorities over the long and short periods. Similarly, this will provide the preconditions for concrete appraisal of the mutual effect on each other of installations above and below the ground.

10.1.2 The town plan

The local authority has a monopoly on planning

The inception of planning obligations regarding use of the subsurface and subsurface installations entails a double obligation. Those responsible for particular functions have the responsibility of presenting the demands for space and installations and of requesting planning. Those responsible for overall planning, represented by the Building Committee, have the obligation of contemplating and considering use of the subsurface and of drawing up a town plan which makes possible the implementation of construction.

In all important respects, the routines in existence today for the drawing-up of proposals for the town plan should be applied. The routines for dealing with the proposal for ratification should also be suitable for plan proposals concerning subsurface installations.

The research project is of the opinion that special demands should be made on descriptions and appraisals of subsurface installations. These can be summarised as follows:

- Overall geological mapping
- Required geological information in detail
- Overall hydro-geological mapping
- Required hydro-geological information in detail.
- Account of the deep foundation of existing above surface installations which are affected by subsurface installation.
- Description of construction techniques to be used in connection with the subsurface space, also initial blasting points and other shafts.
- Description of disposal of surplus excavated material.
- Description of consequences for the surrounding land and buildings with respect to vibration, noise, lowering of the ground water and settling.

It should be possible to carry out appraisals and descriptions of the functional context and content in a similar way for surface and subsurface installations. If the functional content requires examination in accordance with the law on roads, the water law or possibly some other legislation, then the functional examination should be in accordance with these laws. Rock cavities should, however, be assumed to be buildings and as much be subject to the requirements for planning laid down in the Building Law.

The requirement for detail planning in connection with the town plan should not be made more stringent than in the case of above surface projects. If an installation cannot be interpreted as a part of a built-up area then examination as regards location can be carried out in connection with examination for building permission. The requirements for preliminary studies and consequence analyses should, however, be the same as in the case of drawing up proposals for the town plan.

The Building Committee does not as a rule have access to personnel with the necessary special knowledge of subsurface construction matters. In certain communities, few in number, there can be found in other parts of the local authority administration the expertise which the Building Committee may need to make use of. In most communities it is necessary, however, to engage experts on a consultant basis. There should be training of personnel within those communities which are affected or likely to be affected by subsurface construction.

The maintenance of secrecy in the case of subsurface installations requires special care. In most cases the Building Committee as the existing authority can decide on giving secrecy status to information which may reveal an installation or its function. The information which is overt in the town plan proposal and the accompanying documents should be formulated in such general terms that the activity becomes no more apparent than it would be at the construction stage.

10.1.3 Overall planning

The arguments which are applicable for the intensification of overall planning in communities in certain regions and in the country as a whole are extremely relevant in the case of subsurface installations and use of the subsurface. Town planning occurs in the first instance as preparation for impending implementation and is based on planning considerations which have made clear the connection, the choice of alternatives and the consequences both in time and space. Consideration of above surface or subsurface siting of installations and systems should, however, as already stated, occur in the

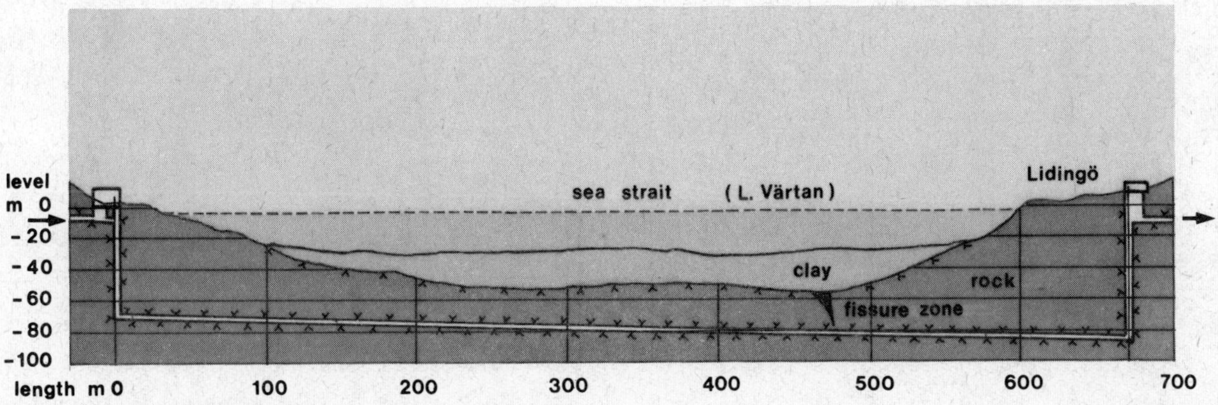

Figure 56

Illustration of how the main tunnel to Käppala is led under a strait

Vattenbyggnadsbyrån

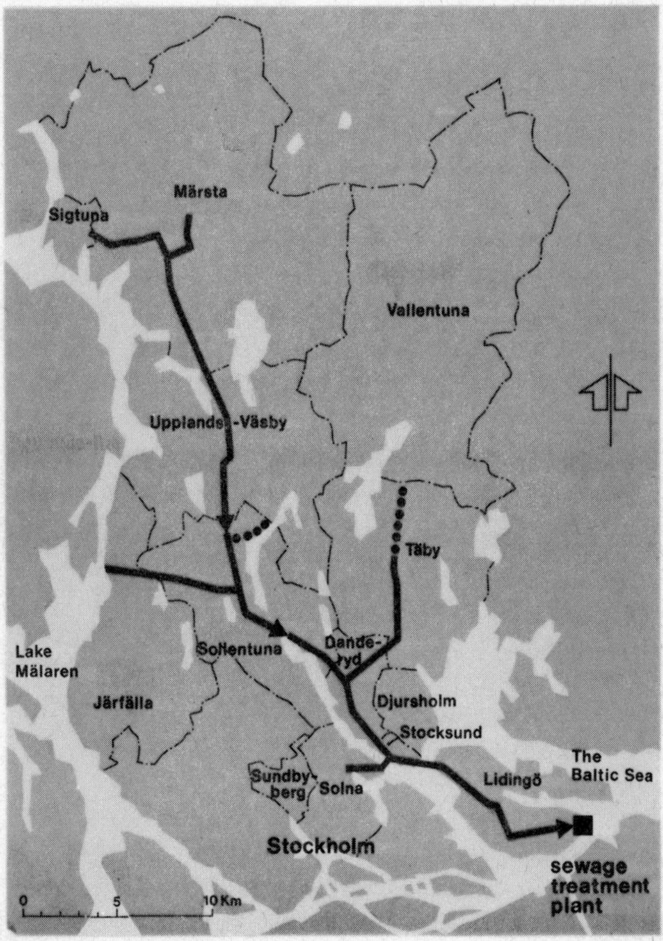

Figure 57

Via a regional tunnel network practically all the waste-water from the northern communities of the Stockholm area is collected into a joint treatment plant (Käppala). The advantages of large scale sewage treatment are combined with the fact that outflow takes place into the most suitable recipient.

Vattenbyggnadsbyrån

partial plan. The connections between use of the subsurface and use of the surface, whether concerned with combination of functions or mutual disturbance, should also be clarified in partial planning.

The inception of planning requirements for subsurface installations will create the conditions in which the various overall planning sections in a local authority will give consideration to use of the subsurface.

The local authority, through its council, is responsible for the overall planning within the community. Those responsible for planning of particular functions are given the obligation of presenting to senior local authority officials long- and short-term programmes for requirements regarding particular functions. These officials, for their part, are given the obligation of initiating examination of utilisation of the subsurface within the committees and administrative departments which supervise the various parts of overall appraisal.

It should be possible as a result of partial and community planning to discuss the requirements for reservation of areas of land for, on the one hand, the careful exploitation of particularly favourable subsurface resources for future needs, and for, on the other hand, the preservation of the freedom to satisfy likely needs. As a result of such planning, one need amongst others should become apparent, namely, the need to introduce regulations to prevent the construction of deep basements and deep foundations which would endanger future construction of tunnels.

It is in overall planning that the geological and hydrogeological mapping which gives important information in connection with utilisation of both the surface and the subsurface should be initiated.

The routines which in various forms of overall planning are most needed concern partly the stimulation of planning of particular functions to work out the required long-term plans for particular functions which take utilisation of the subsurface into consideration, and partly the finding of a suitable division of responsibility in overall appraisal of subsurface installations.

10.1.4 Authority responsibility

The system of regulations in the Building Law demands co-operation between local authority and state in planning at different levels. Dealing with matters concerning subsurface installations and utilisation of the subsurface requires therefore both expertise and authority responsibility at county level and with the central planning authorities.

The question may be raised as to whether special subsurface expertise should not be made available, through staff reinforcements and training, to the county councils, which act as controlling and ratifying authority.

The State Department of Planning should be the authority given authority responsibility for the planning for utilisation of the subsurface and of subsurface installations.

10.2 The right of disposition in the subsurface

10.2.1 The present situation

In principle, a landowner also owns what lies below the surface. The right of ownership thus gives right of disposition of the subsurface. Swedish law does not recognise any stratified right of ownership of real estate property. If a tunnel or other subsurface installation is to be constructed beneath the property of another party, it is thus necessary to obtain special right to do so. In practice, however, the landowner's right of disposition of what lies at greater depth below the surface is limited. The landowner must be able to show that he can make use of

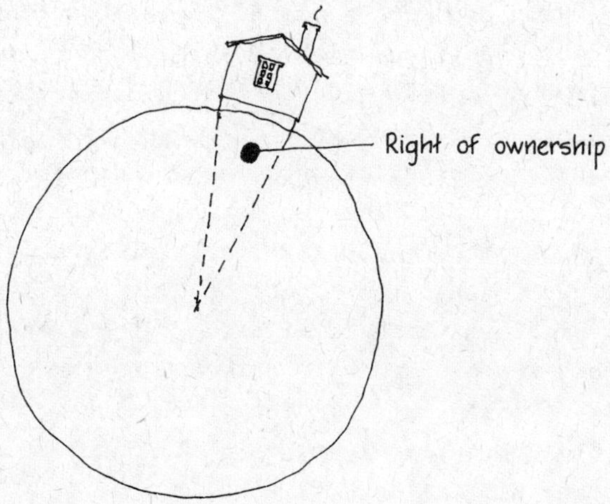

the subsurface in order that a value can be put on it.

The right to excavate a subsurface installation through the property of another party can be obtained in various ways:

An easement agreement, which is the most common form, is reached with the property owner. Easement is bound to the property which the owner of the installation has at his disposal. Entry in the register of landed property can make easement permanent and valid regardless of who owns the properties.

A usufructuary right agreement is reached between two parties. The right of usufruct has a time limit of 25 years in town areas and 50 years in country areas. The right can be transferred so that it applies to a later owner, but this is not the rule as in the case of an easement agreement, the right of usufruct can be strengthened by being entered in the register of landed property.

Society can gain access to the subsurface by use of compulsory powers. Any right obtained by use of compulsory powers has no time limit and gives the owner of an installation a more lasting right than a right obtained through agreement. For this reason formal compulsory purchase is occasionally used where the parties are agreed in advance both that encroachment may occur and on the compensation for this.

The forms taken up hitherto are limited in that they are valid between individual properties and property owners. Where a subsurface installation affects several properties or property owners, agreements must in such case be reached with each and every one of these.

A more subtle form is used in legislation (1974) concerning roads and pipelines, which makes provision for an installation in its entirety to be investigated by an authorised official. The latter law applies in the case of

- telecommunications
- high voltage current
- water
- sewage
- district heating, oil, gas
- other proposes for which concession has been granted

In certain cases, the water law applies and the right of disposition is obtained through the water court.

Normally, property owners grant the right of usufruct without demands for special compensation. Cases in which conflict may arise concern tunnels located near the surface which may encroach on the space for the lowest basement storeys.

There are no established rules for the evaluation of the right of disposition below land belonging to another party. Certain cases in lower courts of law indicate different solutions.

. Compensation for a volume of earth or rock extracted from beneath a property will probably not be awarded. However, if it can be shown that the mere existence of the subsurface installation has reduced the market value of the property, then compensation may be awarded.

. Judgement has been made on the question of compensation for encroachment in some cases, without resultant clarity for the future. The town or city plan restricts the property owner's right of disposition to two, in certain cases three, underground storeys, i.e. 7 to 10 m. If this right of usufruct is not involved, then the possibility of obtaining compensation for encroachment would seem to be slight.

The local authority can through the town plan determine the location of a future tunnel. It can also introduce regulations to limit the right of a property owner to build below a certain level, to give freedom for construction of tunnels, for example. In this way the possibilities for a property owner to prevent the implementation of a subsurface installation and to demand compensation for encroachment are limited.

10.2.2 Desired improvements

The choice of subsurface location reduces the influence of the value of land on a project, and similarly reduces the likelihood of conflicts over use of land. The conditions are thereby created under which property law and associated questions in connection with project implementation can be dealt with for more easily in the case of subsurface location than in the case of surface location. These possibilities for time-saving and simplified procedures should be made use of.

In the new generation of laws, represented by legislation on roads and pipelines, the handling of, for example, the right of disposition, is linked to the project, not to properties. An authorised official consults in a fairly free way with the individual parties and authorities concerned. When agreement between the parties has been reached, the official makes a decision in which questions connected with the creation of the pipeline are specified. The official decides all matters, including the question of compensation for encroachment and other damage.

The present system concerning the right of disposition of land for subsurface use with easement or usufructuary right agreements with the landowner is adequate in most cases. If, however, the landowner proves difficult, this can cause important changes to the time schedule, taking into consideration the short time which there often is between negotiations on agreement and the start of work.

The basic principles of the legislation concerning pipelines, with an authorised official, should to a great extent be able to be applied even in the case of other types of subsurface installation. The authorised official can then also decide whether the proceedings should also involve interested parties other than the owners of overbying land. The present system is limited to those landowners who are directly affected by the subsurface installation. The proceedings should appropriately concern all interested parties within the risk zone.

Legal practice is unclear on several points.

The legislation concerning pipelines has not yet been used to any great extent since its applicability in the case of subsurface installations is not entirely clear. It is intended to apply to all types of pipeline, but the meaning of this is not unambiguous. For example, it would seem that tunnels for stormwater without the requirement of examination for concession fall for the time being outside the law. If, however, a tunnel of this kind is combined with a pipe for wastewater, then the law would seem to apply. Furthermore, larger tunnels causing "encroachment" which is not inconsiderable in relation to the "utility" gained fall outside the law. The meaning of this is unclear.

On what grounds can a property owner refuse to relinguish the right of disposition? He may, for example, not like the use to which the subsurface installation is intended to be put. The question is of interest for both parties.

A usufructuary right agreement according to the land laws is, as already stated, normally of limited duration. Considering the fact that rock cavities are long-lasting, a right without time limit would be preferable.

10.3 The right to construct subsurface cavities

10.3.1 Valid regulations

The first paragraph of the Building Law is as follows:

"The use of land for building on shall to the extent prescribed in this law be preceded by planning. Permission is required for building (building permission) in the scope decided by the government."

In accordance with paragraph 54 of the building statutes, building permission is required for the following:

. new building
. alteration of bearing structures, fireplaces, smoke or ventilation ducts
. installation of pipes for water supply or sewerage
. erection or important alteration to walls and planks
. excavation or filling which means considerable change in the height of the site
. installation of fixed cisterns or other fixed arrangement for handling or storing inflammable liquids over and above the household requirements of the property
. installation of support

A subsurface installation which does not fall under the above definitions can thus be constructed without building permission. Building permission may, however, be required if the subsurface installation is provided with for example sewerage facilities or if the installation is to be used for storage of inflammable liquids such as oil. There is comprehensive legislation which concerns the objectives of work carried out in subsurface installations. Examples of such legislation are the following:

. The water law controls changes to the catchment area and its runoff.
. The Environment Protection Act deals with discharge that may lead to pollution.
. The law on electrical installations.
. Legislation on roads.

Stormwater tunnels, for example, and the effects of these on recipients are subject to comprehensive technical and legal procedures. The most important feature of this special legislation is, however, that it only deals with the nature of the activity. The structural design of the cavity itself is not subjected to any sizeable consideration.

With the exception of excavation and other work at ground level which is required for initial blasting etc., it is in other words possible for large subsurface installations to be constructed without any scrutiny of plans or construction techniques, let alone any inspection on the part of the Building Committee. This may be regarded as remarkable. In the case of an above surface project, the Building Committee is to carry out

. an examination of the documents concerning building permission and a check that safety is satisfactory from the points of view of strength and stability; a check of insulation against water, damp, cold and disturbing noise etc.;
. an examination of design calculations, covering load assumptions, methods of calculation and allowable stress;
. an inspection of building work on various occasions during execution;
. an inspection of the completed building;
. collection of experience and services of an advisory nature.

The state and county authorities are in principle exempt from the requirement of obtaining building permission, and it is sufficient that they notify the Building Committee.

Subsurface installations, however, are not entirely exempt from controls. There is a certain amount of control exercised by the Industrial Safety Board in accordance with their directive No. 33: "Directives concerning inspection of cavity conditions." According to these directives:

. rock cavities are to be subjected to official inspection and scaling (removal of loose rock); free-standing structures built into the rock cavity shall be so designed that the rock can be inspected;

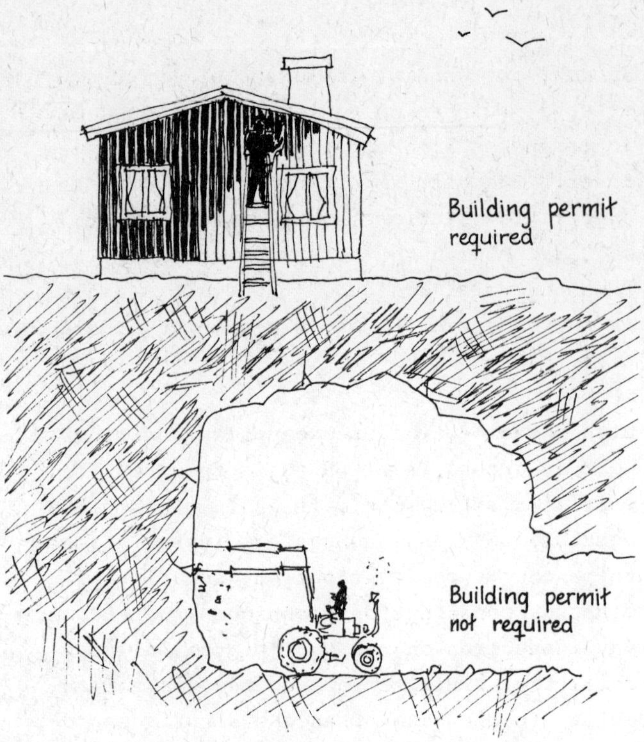

10.3.2 Experience

. before the installation comes into operation, scaling and possible reinforcement or support work is to be subjected to careful official inspection;
. Scraping is to be carried out at various intervals even after the rock cavity has come into operation.

It is characteristic of these directives that they are from the standpoint of industrial safety. Inspections take place at the place of work. The Labour Inspectorate and the Industrial Safety Board are not accorded the power to examine the design of a subsurface installation. This shall, however, be in line with the applicable sections of the directives on rock issued by the Industrial Safety Board (No. 67). During the construction stage, subsurface installations are visited by the Labour Inspectorate's mines inspector, who may then direct, for example, reinforcement measures over and above those already designed for or carried out. In consultation with the client and the contractor, he also decides progressively on periodec official inspection and post-scraping (incl. supervision).

In a number of cases subsurface installations have been subjected to examination for building permission. Local authorities with extensive subsurface construction work have considered this examination procedure to be of value. In a community, where extensive oil storage installations in rock had been planned, the Building Committee extended its sphere of authority. Following agreement with the responsible party, the Building Committee has examined and closely followed the construction work. It would be noted that even in those cases where the Building Committee is local authority has attempted to scrutinize or to participate in subsurface construction, this has concerned individual projects. For example, different courses of action have been taken depending on whether construction has been under the auspices of the local authority or not.

It can be stated in summary that society does not possess the means to exert influence on subsurface construction - apart from the few exceptions mentioned above. The major part of subsurface construction which is executed in Sweden takes place without the pre-examination and follow-up which applies in the case of construction above surface. The qualitative level achieved in the execution of the project depends on the degree of competence of the responsible party and those he employs.

Particularly in the case of local authorities, these have in most cases not been given cause to increase their competence within problem areas associated with subsurface construction - local authority officials have expressed this point in several cases.

10.3.3 Examination in connection with building permission - desired level

General

Physical planning above the surface is accompanied by a form of control in which society (the Building Committee) examines

- whether the proposed project is in line with plans currently in force
- whether the project has been preceded by studies of the kind demanded in connection with safety to human beings and property
- whether the execution of the project conflicts with laws currently in force and with directives based on these laws.

The parties affected by the project are given the opportunity of expressing their opinions. Furthermore, inspection of the construction work and of the completed project is carried out. Collection of experience and services of an advisory nature make an active part of the work of the Building Committee, in addition to its function as a controlling body.

Through the procedure for building permission, society is guaranteed knowledge of building activities and is provided with the possibility of influencing both planning and construction. The argument for the participation of society in these matters is valid whether construction takes place above or below the surface. Subsurface acticity has assumed such proportions that there is justification for interest in it on the part of society.

Before formulation of the requirements that should be met in examination for building permission of subsurface installations, a number of particular conditions should be pointed out:

- The long-term nature of rock cavities. Cavities should be so designed that the stability of the rock is not changed with time.
- Safety factors for human beings and property during both the construction and operation stages make special demands.
- Special attention must be paid to the effects on the environment in the case of installation below the (upper) groundwater level.
- The need for co-ordination between different subsurface projects and with existing subsurface installations demands attention, with consideration to conservancy of nature resources in the subsurface, to the difficulties of orientation in three dimensions, and to demands on stability of the rock mass.
- Different factors, such as type of installation, its location, geological and hydrological conditions etc. can exhibit wide variation. The requirements for examination for building permission should be adaptable to the conditions prevailing in each individual project, so that the owner and the authorities are not burdened, but rather that there are corresponding resultant benefits.

The scope of examination for building permission

On the basis of what has been mentioned above and in other parts of the report, examination for building permission for subsurface installations should happen with the following general guidelines:

- A check that the installation does not conflict with plans currently in force nor with future planning intentions. Examination of the possibilities of combination with other projects.
- A check on the right of disposition, which where applicable, is supported by completed proceedings.

A check from the construction point of view. The inquiry shall, where applicable, confirm the following conditions:

- geological overview
- hydrogeological overview, based among other things on ground-water observations over a period of at least a year.
- account of existing buildings, roads, pipelines/cables, with method of foundation, within the area which can be affected by ground-water drawdown (area of influence)
- account of position of rock surface
- forecast of necessary reinforcement measures, with reference to stability, safety and impervionsness
- account of programme for control measures
- account of calculations on necessary protective and risk zones

. Checking from the functional point of view. The intention here is to check that the operation environment and safety fulfill the requirements.

. Checking that the relevant regulations which in other legislation control the activity in question are respected.

. Consultation and checking in connection with disposal of surplus excavated material.

An account is given of the installation, together with working documents. Inspection from the point of view of stability and safety can best be carried out in co-operation with the Labour Inspectorate. Inspection with respect to effects on the environment should take place in consultation with experienced hydrogeologists and where necessary, also with experts in ecological questions. The national body of experts which is proposed elsewhere (see section xxx) should when necessary be able to recommend hydrogeological and ecological expertise.

It is assumed with the granting of building permission that

. geological mapping of the cavity is carried out continuously throughout the construction period

. geological mapping is accompanied by mapping of the reinforcement and sealing work carried out

. the completed cavity is documented with regard to position and layout (report document)

It is assumed that control measures are carried out before, during and after construction work to the extent determined in consultation with the Building Committee, specialists engaged by the Committee, and the Labour Inspectorate.

Documents shall be submitted to the Building Committee with energy application for building permission. These are:

. completed preliminary investigations

. analysis of the risk of effects on the environment

. working documents

. plan for control measures

On completion of the installation, the following documents are to be submitted to the Building Committee:

. account of control measures carried out and the results of these

. report document

. geological mapping of the cavity and mapping of the reinforcement and sealing measures carried out

In certain communities, there are local authority departments with extensive subsurface activites which command their own subsurface expertise. In communities of this kind, the Building Committee should engage this expertise for the special technical examination and registration. Not included in this, however, are controls and considerations of a planning nature. The documents mentioned above should in all circumstances be submitted to the Building Committee.

Figure 58

Cavern storage of oil products is often less costly than surface tanks. The difference becomes significant at larger volumes and/or higher land prices. The 500,000 m³ plant for heavy fuel oil at Oxelösund, shown above, is one of more than 200 plants currently in use in Sweden.

Skanska

The state and the conty councils should not be granted exemption from the obligation to apply for building permission.

For some kinds of activity, examination may occur in accordance with some law other than the Building Law, for example, the laws on roads and water. Consultation with the Building Committee should occur in such cases.

10.3.4 Viewpoints on the safety aspect

Safety with regard to the people and activities in a subsurface installation requires careful consideration with respect both to the layout of the rock cavity and to the equipment, as well as to the work routines applied during the utilisation stage.

The functions for which the installation is intended must be carefully studied. The presence of, for example, high tension current, hot water, gas, storm water, waste water or motor vehicles in subsurface installations makes up a special risk factor. For example, the failure of a hot-water pipeline would have a completely different effect in a tunnel to the effect of a failure above surface. The pipeline concerned would have to be fitted with an automatic shut-off device which would also have to be located bearing in mind the length profile of the tunnel. A combination of risk factors requires special consideration. For example, when are high tension cables and gas mains compatible? Similarly, hot water pipelines and motor vehicles?

Escape routes must be dimensioned with respect to quantity and capacity not only in relation to the intended function of the tunnel, but also to suit demands made during the construction stage. A storm water tunnel which can suddenly fill with water in the event of a storm must have adequate escape facilities if human beings are to be allowed access to them at all. (There are also cleaning and inspection requirements). If the nature of the activities carried in a subsurface installation demands it, the escape routes must be dimensioned to allow rescue vehicle access.

An extreme example of the implications of the fact that subsurface construction does not come under the jurisdiction of the Building Law or within the area of responsibility of other authorities can be seen in the case of the construction of the Stockholm metro. When the main party involved, AB Storstockholms Lokaltrafik (the Stockholm Passenger Transport Board), discussed escape routes from metro stations with the authorities concerned, it was found that there were no applicable regulations included in the Swedish Building Standards for the very simple reason that metro was not defined as a building. Following consultation with the Swedish Board of Urban Planning and the fire authorities in the communities involved, a solution was reached concerning the stations themselves, whereas the rest of the tunnel was, and still is, outside the jurisdiction of the authorities concerned. (The metro carries about 400 000 passengers daily).

The design of fresh-air intakes and ventilation facilities should be subject to special demands on safety and dependability. In the USA, a serious accident occurred when chlorine gas leaking in through an air intake, spread through a metro system. In a similar way, severe accidents have occurred when fires and the resulting toxic gases have been spread within subsurface installations via ventilation systems.

The fact that the subsurface is completely dark makes demands on lighting arrangements. If the normal lighting system should fail for some reason, then there must be a simple procedure for emergency lighting. In each individual case, the safety arrangements should be tested with respect to various emergency situations. One advisable way is to arrange simulated situations where the safety systems and human beings are tested in special circumstances. Analysis must not be limited to the actual installation itself, but should also include community services such as the fire brigade and ambulance service, any special equipment that may be needed, access routes, etc.

People who are to spend their time in a subsurface instllation should be instructed on matters of safety before being admitted. This is difficult to do in metros, shopping centres, and similar public premises. Instead, safety must be incorporated in the installation itself.

10.4 Control measures

10.4.1 During the construction stage

A certain amount of control of installation work is today carried out by the Labor Inspectorate. This control is linked to industrial safety (see section 5.1.2). If examination for building permission were introduced, this would mean that responsability for control by society from other points of view would fall on the Building Committee.

The Building Committee's control measures would in this way be concerned partly with supervision of the layout of the installation, and partly with foreseeing and reducing the effects of possible damage. It should be in the owner's interest if such measures were to come into being.

As a rule, the following measures for prevention of damage and for registration should be taken:

- Measurements of ground-water level are carried out several years before, during and after construction work. Water which has leaked into the installation is measured and recorded by means of overfalls and pumping.
- Measurement of settling is carried out in the same way as for buildings and installations that can be expected to settle.
- Measurements of vibration are carried out in buildings and installations which can be expected to be damaged by vibration and shaking.
- Buildings and installations are inspected before and after blasting work.

Control measures must be extended over a sufficiently wide area around the installation, which is not always the case today.

With respect to the fact that ground settling develops over long periods, due consideration should be given to the fact that the guarantee period for subsurface installations should be made longer than for other types of building.

There are today available directives for installation work which have been worked out by the Industrial Safety Board, and partly by the Fortifications Administration. Control on the part of the Labour Inspectorate aims at protecting those working in an installation during the constructing period. The client often exercises of the contractor's work, with the aim of ensuring execution of work in accordance with the contract.

None of these control measures is motivated by a responsibility to safeguard the interest of the general public as far as operation or use of an installation is concerned, nor to watch for risks of effects on the environment which may concern either society in general or private individuals.

Control carried out by the Building Committee should above all be concerned with the latter. It is, however, desirable that control be carried out in co-ordination with that control carried out by other interested parties.

10.4.2 Measures in connection with operation and closure

Control measures with regard to effects on the environment have been referred to above. Depending on the type of activity, subsurface installations are subjected to stress in the form of vibration, leakage of substances that can effect the rock material, internal and external stress, etc. Rock which has been drained in connection with excavation can be subjected to deterioration processes in the form of weathering, conversion to clay, etc., as a result of oxidation processes following the replacement of ground water by air. Geological plotting of rock cavities, with the corresponding plotting of reinforcement carried out, is an extremely valuable aid in the appraisal of risks and of the need for action.

Installations which are no longer in use today often closed off by sealing up the entrances or by similar measures. This means that no control is carried out concerning changes that may occur in the rock cavity.

In view of the prevailing shortage of documentation and registration, there is risk that installations are completely forgotten or at least that very little is known about their extent. If an installation of this type fills with water, which occurs if it is located below normal ground-water levels, it constitutes considerable risk factor for future subsurface installations built in the vicinity and at a lower level. In the event of any form of hydraulic communication, break-through or other distribution system, instant flooding of the new installation would occur.

11. RESPONSIBILITY AND DAMAGES

It must be assumed in the case of a well-functioning subsurface installation that problems associated with responsibility and compensation for damage will be settled in a way which is satisfactory to all parties concerned. Legislation must naturally be applied in accordance with the same principles regardless of whether building work is being carried out above or below ground level.

Certain types of damage which can be caused by construction work above ground and also by subsurface construction work are of a particularly problematic nature, for example, as far as determination of the cause of the damage and the causal relationship are concerned. As a result of this, it is desirable for the application of legislation to be as simple and unambiquous as possible.

11.1 A description of the situation today

In the construction and utilisation of subsurface installations, damage can occur to neighbouring buildings or to buildings located further away. Damage can primarily be caused by

- vibration and shaking
- settling resulting from a lowering of the ground-water level

Damage caused by shaking or vibration

Damage caused by vibration and shaking is covered in detail in current legislation with respect to the Swedish Environmental Protection Law and the Swedish Land Laws.

Responsibility is in principle clearly defined and damages are awarded even if the person who has carried out the work has not been guilty of negligence.

The strict responsibility rules are specified as follows in the Swedish Environmental Protection Law.

§ 1.

"The use of land, buildings or facilities in a way which can cause disturbance and interruption to the surroundings through air pollution, noise, shaking, light or other forms of interruption which are not purely temporary."

§ 30.

"Those who cause inconvenience through environmentally dangerous activities are to provide compensation for such inconvenience. If the inconvenience has not been caused by carelessness, however, compensation is to be paid only if the inconvenience is of some significance and only to the extent that it cannot reasonably be tolerated with respect to prevailing conditions in the district or to its general occurrence under comparable conditions."

The Swedish Land Laws specify the following with respect to legal conditions between neighbours:

§ 4.

"If digging or similar work is particularly extensive or implies a particular risk to the surroundings for some other reason, then the damage resulting from the work is to be compensated for even if no negligence has been committed."

The Swedish Environmental Protection Law specifies strict responsibility if the disturbance is not of a purely provisional nature and if the specifications of the Law are not satisfied. The result of this is clearly defined responsibility in connection with the blasting of subsurface installations. The Swedish Land Laws assume that the work is to be particularly extensive or is to imply special risk for the surroundings if strict responsibility is to apply.

On condition that the owner of the building is aware of the fact that a subsurface installation is being built, all he has to do is to show that damage has occurred as a result of and in connection with such subsurface work in order to obtain compensation. The strict responsibility is not limited to the neighbours alone but also applies to the owners of other buildings close to subsurface installations.

Settling damage resulting from lowering of the ground-water level

There are no regulations for this type of damage in Swedish Law. Furthermore, there are no cases from the Supreme Court which can clarify the way in which such damage is to be treated. This means that settling damage resulting from a lowering of the ground-water level becomes considerably more complicated to appraise.

The Swedish Water Law includes certain regulations concerning damage when someone utilises the groundwater or damages the groundwater belonging to another person. Permission is required from the Swedish Riparian Court to utilise an amount of water which exceeds 300 m^3/24 hours. This corresponds to about 3.5 litres/second. Such a large quantity of water is seldom left uncontrolled at one and the same leakage point in our subsurface installations. Criteria up to now are less than or equal to 1 litre/second and kilometer of tunnel, and it would be necessary to have a tunnel about 3.5 km long in order to reach the limiting value specified by the Swedish Water Law.

The strict regulations concerning damages in connection with the Swedish Environmental Protection Law and the Swedish Land Laws would also appear to cover settling damage resulting from a lowering of the groundwater level. At the moment, however, we have no practical legal experience in this field. In one major dispute, the parties concerned reached a compromise under conditions which imply that the builder accepted at least a certain amount of responsibility in this respect. The problems involved in settling damage are complicated. Apart from the purely legal problem, it is technically difficult to determine whether settling has occurred as a result of filling processes in the ground and/or a general change in ground-water level within the area or whether this settling has resulted from a lowering of the ground-water level caused by leakage to subsurface installations. Depending on the geological and the hydro-geological conditions, as already mentioned, extremely slight leakage can have disastrous consequences on the ground-water pressure level. In areas consisting of loose and compressible clay, the balance of the ground-water pressure level is controlled and even a particularly slight leakage of the order of less than 1 litre per second and kilometer tunnel can result in damage. Thorough preliminary surveys can clarify whether there is risk or considerable risk of settling damage. These surveys, described in Section 2, are, however, both time-consuming and expensive and can normally not be carried out by a person who has suffered damage.

The legal security of a third party

Damage associated with the lowering of the ground-water level is often comprehensive from the viewpoint of buildings and can affect property which has no direct association with the party responsible. The possibility of a third party avoiding damage or, in the event of damage occurring, of obtaining reasonable compensation for it becomes complicated for several reasons.

1. The difficulty in anticipating the risk of damage. Today no demands are made on either planning work or building permission. Contrary to the case of building work carried out above ground level, a future subsurface installation does not need to be preceded by a detailed account of the plan, does not need to be put on public display and need not attract any attention whatsoever. (If the installation concerned passes through the zone of ownership of the person owning the property, then the construction work has been preceded by a settlement of right of disposition. Neighbouring buildings, on the other hand, are not affected.) As has been shown on several occasions when damage has occurred, the owners of property can be totally unaware of the fact that subsurface installations are being constructed within the hydrogeological zone of influence.

   In cases where the owner of property finds out about a planned installation - and in such good time that the project work can be influenced - it is still necessary for him to have considerable knowledge of general and local geological conditions, if he is to be able to appraise whether there is any possible risk of damage.

2. The burden of evidence - to show relationship between cause and damage

   In order to prove his rights, a person suffering damage must be able to show that:

   - a subsurface installation (for example) has caused a lowering of the ground-water level
   - the lowering of the ground-water level has resulted in settling in the ground
   - the settling in the ground has caused the damage

   As shown in section 9.3.3 , this causal relationship is generally complicated and more than one factor can exert its influence. In order to be able to demonstrate a lowering of the ground-water level, it is necessary for the prevailing conditions before the creation of the installation to be known, that is to say, before the damage occurred. Surveys of this type are time-consuming and expensive.

Figure 59

When settling becomes as extensive as shown in the picture, ingoing services are cut off, there is a risk of piles cracking and the house being completely destroyed.

STEGA

3. Practical legal experience in these matters is not at all extensive. One complication is the fact that matters concerned with the Swedish Environmental Protection Law are considered by a special court, the Property Court, while the Swedish Land Laws and laws concerning damages are handled by a general court of law.

4. At the present time, the owner of property cannot obtain insurance cover for damage resulting from vibration and shaking from blasting or settling caused by a lowering of the ground-water level. Support from experts at insurance companies is not normally available since these experts often represent the responsibility of the party carrying out the construction work.

5. Different types of survey carried out by authorities can be kept secret on the basis of the paragraph in the Swedish Official Secrets Act concerned with internal documents. The surveys never need to reach any other status. In this connection, for example, general distribution of information which is valuable from the point of view of both responsibility and questions of damage in connection with subsurface installations is prevented. The most flagrant example found by the survey in this connection concerns a tunnel which resulted in a lowering of the ground-water level. Persons not involved have been permitted to study part of the individual ground-water level measurements but have not been allowed to study a composition drawing made which clearly showed local lowering of the ground-water level. The motive for secrecy was stated as being that "if a map of this type became known, the holder would be immersed by claims for damages and this would be unreasonable".

Summary

To date, damage caused by subsurface installations have been compensated for by amounts which have added up to a total of tens of millions of Swedish Crowns. The extent of the damage which has occurred but has not been compensated for is difficult to appraise, but can be expected to attain an equivalent of even larger sums of money.

The difficulty for a person suffering damage to reach a position in which compensation is awarded has been described above. In combination with the general anonymity in society which has hitherto characterised subsurface activities, this fact can in many cases be assumed to have persuaded the main contractor or his representatives to act without taking sufficient caution or to take risks. This concerns cases where insufficient insight into the relationship of cause and effect is not itself a sufficiently good explanation. This comment is made necessary by individual observations in connection with surveys carried out and does not refer to the branch as such.

11.2 The need for reforms

It is suggested in this report that legislation be introduced in which requirements are specified whereby subsurface installations should be preceded by planning, should be submitted for consideration in connection with building permission and should also be subjected to technical construction control. The demands should be accompanied both by standards and by instructions. The legal situation regarding responsibility and compensation for damage would be improved if clear instructions were supplied by the National Board of Urban Planning to the various building committees to the affect that analysis of effects is to be carried out to provide guide-lines for decisions with respect to the planning and construction of subsurface installations.

It would appear that legal security for a third party could be included in the strict responsibility regulations suggested in the Swedish Environmental Protection Law and the Swedish Land Laws. Today it is too early to decide how the highest legal instances will apply these new laws in the case of damage resulting from a lowering of the ground-water level. The general interpretation among experts in this field, however, is that compensation will be paid for damage which has its cause associated with subsurface installations. It appears important for measures to be taken to develop unambiguous and easily applicable legislative interpretation in this field.

The limited protection provided by normal property insurance does not cover damage which can occur in connection with the building and operation of subsurface installations. This means that the owner of the property must demand compensation in the form of damages. Few property-owners are able on their own to follow the technical and legal surveys necessary in such cases, surveys which are often of a very extensive nature. Better balance would be achieved if the property-owner (the party suffering damage) could be represented in a more active way by experts in the insurance companies, either through an extension of legal protection insurance associated with normal property insurance, or by a supplement to property insurance providing protection from damage caused by vibration, shaking and settling.

12. WHAT SORT OF REGULATIONS ARE REQUIRED?

In the report on general principles, entitled "Utilisation of land and construction", which is based on the findings of committee of inquiry into the building laws, it is pointed out that subsurface construction is causing increasing amounts of damage to buildings, traffic arteries and pipelines and cables jus below the surface.

The conclusion reached in the inquiry is that conditions should be improved through "expedient regulations concerning planning and technical inspection in connection with subsurface construction and foundation work."

Undoubtedly Swedish building legislation is already so comprehensive that further supplementary legislation would probably make conditions worse rather than improve them. Instead, conditions could probably be considerably improved if, as the committee of inquiry put it, expedient regulations were drawn up.

These regulations should include directives for the planning stage, information on problem areas and precise instructions concerning the conditions under which building permission can be granted.

Important changes in the Building Law should, however, be made in order to achieve the desired degree of involvement by society in subsurface construction and utilisation of the subsurface. Such changes would define subsurface cavities in the form of tunnels and chambers in rock or other material as buildings.

These changes should result in subsurface utilisation and subsurface installations being subject to planning requirements and demands for examination for the granting of building permission. The directives should contain detailed information concerning the requisite obligation to render accounts in connection with various types of planning and in applications for building permits.

The directives should also provide information regarding:

a) Demands for inspection in connection with the construction and operation of subsurface installations.
b) Demands for the registration of subsurface installations.
c) Demands for the registration of geological and hydrogeological information.
d) Demands that the companies and national and regional bodies responsible for particular functions work out programmes and plans where claims for utilisation of the subsurface are given the same time horizons as those applied in local authority planning.

For the administrative management of plans and projects which involve subsurface location, those authorities which currently deal with or are active within the field of subsurface construction and planning - county councils, the Department of Planning, the Nature Conservancy Department, the Industrial Safety Board and the Social Welfare Board - should be given official responsibility for subsurface construction in the directives.

The local authority should, through directives, be instructed to appoint committees and administrations which are to be responsible for the measurement and registration of subsurface installations and geological and hydrogeological information.

It can be debated whether or not the Department of Planning, pending changes in the law, should draw up recommendations for local authorities and county councils to the effect that subsurface utilisation and subsurface projects be included in local authority planning and that demands be made for building permission to be compulsory for subsurface installations. It is impossible that such a recommendation would conflict with the laws currently in force.

13. HOW CAN KNOWLEDGE BE SPREAD AND DEVELOPED?

Information

Work on research tasks has revealed a considerable lack of knowledge regarding subsurface utilisation and subsurface construction at all levels, both within companies and local authorities and on the part of government bodies and authorities.

The intention of the measures recommended above that administrative routines and regulations be introduced through legislation and directives is to make demands for a quantitative increase in planning and design work. In this way it assumed that the amount at attention focused on subsurface utilisation and subsurface construction will increase.

A qualitative increase in planning and design work will only occur through the spreading of existing knowledge via information and training and through research and development work. The information should aim at making the possibilities techniques and problems attached to subsurface construction more widely known among specialists, administrators, decision-makers and the general public.

The work in connection with information must be divided into several individual efforts: the collection of facts and experience, the processing of information for different categories of recipient and the strategy to be employed for spreading the information.

Parties interested in the information include consumer categories, industrial plants, people responsible for research and development within the field as well as society as being responsible for utilisation of resources, risk evaluation and co-ordination.

Work in connection with information has up to now been directed towards more or less specialised individual aspects. There is no body or institution which at present has either the resources or the responsibility for carrying out the requisite work.

Training

The planning of subsurface utilisation and construction of subsurface installations demand the co-ordination of a great number of specialist skills. Three groups of specialist fields can be distinguished:

the construction of cavities

the adaption of functions to subsurface cavities

the co-ordination of cavities and subsurface functions with the surroundings

There is at present no form of training which covers all three fields. It is probably also reasonable to assume that in the future there will be co-operation between groups of specialists with different types of basic training.

What is necessary, however, is for supplementary training to be introduced within the various specialised fields, which partly make possible specialisation in subsurface construction, but also provide a basis for co-operation between specialists in the subsurface field.

The construction of cavities is covered by the specialist spheres of geo-knowledge and installation technique. Training in this context has been through practical work leading up to specialisation in subsurface construction. An overall term for the field is Engineering Geology. A recently started form of training within this specialised field and closely related to mining engineering is offered at the technical college in Luleå. A demand should be made for comparable training to be introduced at other technical colleges.

The adaption of functions to subsurface cavities demands that specialists with training within various fields of function be given the possibility of taking part in supplementary training in subsurface construction and in the adaption of the functions to subsurface conditions. Since many functions can be considered for subsurface location training should encompass a large number of specialised technical fields. Consideration should be given both to additional in-service training and to supplementary training at technical colleges.

The co-ordination of cavities and subsurface functions with the surroundings is a specialised field which is synonymous with general and physical community planning. Training concerning subsurface utilisation for these categories is totally lacking at present.

In order to satisfy the demand for training, in-service courses should be arranged and supplementary training be offered at technical colleges or other educational establishments.

Research

Research and development work within the subsurface field is carried out at technical colleges under the auspices of BeFo, the Swedish Rock Mechanics Research Foundation. BFR and STU support this research work and the spontaneous or organised research which is carried out by researchers and associates by providing initiatives and financial help. The Committee for Subsurface Construction within IVA is responsible for stimulating certain forms of important research but lack their own resources.

The present research project has established important areas of research and development such as:

- appraisal of needs
- the collection of data and experience for cost-benefit analysis
- the reaction of human beings to the subsurface environment

These areas of research and development demand continuous collection of data and data processing and should, bearing this in mind, be centred on an institution with permanent staff and annual grants.

Sweden's participation in ITA (the International Tunnelling Association) has led to involvement in international development projects including, among others, the project Subsurface Planning. The programme for this project takes in the following sub-headings

Characteristics of subsurface space
Human reactions during underground stay

Factors affecting subsurface use
Methods in planning

Even this type of development work would be well-served by having access to an institution with a permanent secretariate for international contact.

Central institution for subsurface construction

There is good reason to consider whether or not subsurface construction and subsurface utilisation has reached such a scale and importance in Sweden that a central institution should be given the task of dealing with questions within this subject area.

There is a need for a body of experts who can draw up directives concerning registration of subsurface installations and geo-information as well as administration of subsurface questions in local authorities and can adapt such information to directing the development of subsurface activities.

There is also a need for a body which may be consulted for comment, which is competent to support local and central government authorities in evaluating subsurface problems.

Information, training, research and international contact are other areas of work which demand continuous attention.

Industrial plants and government research bodies have up to now co-operated in the mutual task of developing the technique of rock construction.

The present and future widespread use of the subsurface represents utilisation of natural resources in the service of society. It therefore rests with society to examine how the highly developed techniques for exploiting the subsurface can best be utilised for future demands.

14. REFERENCES

Alternatives in Energy Conservation: The Use of Earth Covered Buildings. Proceedings and Notes of A Conference Held in Ft. Worth, Texas, July 9-12, 1975. (U.S. Government Printing Office.) Washington, D.C.

Berganvisningar, 1975. (Liber Tryck.) Arbetarskyddsstyrelsens anvisningar nr 67, utfärdade 1969, rev. 1974. Stockholm

Bergman, Sten G.A. & Larsson, Torbjörn, 1973, Svenskt tunnelbygge. (Svensk Byggtjänst.) Byggforskningens informationsblad B 10:1973. Stockholm

Brandförsvar under jord. (Sv. Gruvföreningen.) Stockholm

Brännfors, S. et al, 1973, Bergsprängningsteknik. (Norstedt & Söner.) Stockholm (2. uppl.

Buller. (Sv. Gruvföreningen.) Stockholm

Collins, Belinda Lowenhaupt, 1975, Windows and People: A Literature Survey. (U.S. Government Printing Office.) U.S. Dept. of Commerce, National Bureau of Standards, Building Science Series 70. Washington, D.C.

Forsman, Sven et al, 1961, Health Problems in Underground Factories. (U.S. Executive Committee of the 13th International Congress on Occupational Health 1960.)

Halén, Martin, 1968, Ergonometri vid anläggningsarbeten under jord. (Byggnadsindustrins arbetsforskningsstiftelse.) Byggnadsindustrins forskningsrapporter och uppsatser Nr 7 1968. Stockholm

Jansson, Birger, 1976, Draft Programmes For Studies Within The Field of Subsurface Planning. (International Tunnelling Association, ITA.)

Jansson, Birger, 1974, Gå under jord-internationellt perspektiv. (AB Byggmästarens Förlag.) Arkitektur Nr 1 jan. 1974 p. 4-5. Stockholm

Jansson, Birger, 1974, Terraspace - A Hidden Resource. (The Institution of Engineers, Australia.) National Conference Publ. No. 74/5. Melbourne

Jansson, Birger, 1976, Terraspace - A World To Explore. (Pergamon Press.) Underground Space Vol. I No. 1 May-June 1976 p. 9-18. Elmsford N.Y.

Jansson, Birger, 1974, Underground Forecast - Human Ways Out Of An Urbanization Dilemma. (Morgan-Grampian Ltd.) Tunnels & Tunnelling Vol. 6 No. 1 Jan-Feb 1974 p. 30-33. London

Jansson, Birger, 1976, The Use Of Terraspace To Meet Social Demands. Paper submitted at the Swedish Underground Construction Mission to USA Oct. 4-15, 1976. (UDC-Underground Design Consultants, VBB Vattenbyggnadsbyrån.) Göteborg

Legal, Economic, and Energy Considerations in the Use of Underground Space, 1974. (National Academy of Sciences.) Washington

Lindskoug, N-E & Nilsson, L-Y, 1974, Grundvatten och byggande, STEGAs arbete 1966-73.

(Svensk Byggtjänst.) Byggforskningen rapport R 20:1974. Stockholm

Lysén, L. & Palmqvist, K., 1976, Tätning av bergtunnlar. Projekteringsmetodik, injektering och förstärkning. (Sv. Byggtjänst.) Byggforskningen rapport R 4:1976. Stockholm.

Minneapolis Campus, University of Minnesota. I: Potential Use Of Underground Space, and II: Preliminary Design Information For Underground Space, 1975. (Dept. of Civil and Mineral Engineering, University of Minnesota). Minneapolis

Morfeldt, Carl-Olof, 1967, Bergrum och tunnlar. Byggnadsgeologisk uppföljning och kartering. (Sv. Byggtjänst.) Byggforskningen rapport R 15:1967. Stockholm.

Morfeldt, C-O, Nordin, P-O & Roosaar, H, 1967, Byggnadsgeologi. Bebyggelseplanering med hänsyn till den geologiska jämvikten. (AB Byggmästarens Förlag.) Byggmästaren, nr 6 juni 1967, p. 245-267. Stockholm

Report of the Interior Study Mission on Rock Tunneling and Rapid Excavation in Sweden, 1975. (U.S. Dept. of the Interior.)

Rock Siting Of Nuclear Power Plants From A Reactor Safety Standpoint. Final Report, 1975. (Centrala Driftledningen, CDL.) Stockholm

Stauffer, Truman, 1972, Guidebook to the Occupance and Use of Underground Space in the Greater Kansas City Area. (Dept. of Geology-Geography, University of Missouri.) Geographic Publ. No. 1. Kansas City, Missouri

Stockholms tunnelbanor 1975. (Stockholms läns landsting.) Stockholm

Svensk Byggnorm SBN 1975. (Statens Planverk. Liber Förlag). Stockholm

Swedish Underground Construction Mission to the United States of America Oct. 4-15, 1976. (Sw. Export Council.) Stockholm

Tillsyn av bergrums bestånd. (Liber Förlag.) Arbetarskyddsstyrelsens anvisningar nr 33. Vällingby.

Underground Space, 1976. (Pergamon Press.) Vol. 1 No. 2 July/August 1976. Elmsford N.Y.

Undermarksplanering. Lägesrapporter etapp 1, 2 och 3, 1974. Slutrapport etapp 3, 1975. (Statens Råd för Byggnadsforskning.) BFR projekt 730591-8. Stockholm

The Use of Underground Space To Achieve National Goals, 1972. (American Society of Civil Engineers.) New York.

Utnyttjar vi underjorden rationellt?, 1973. (Ingeniörsvetenskapsakademien) IVA rapport 56. Stockholm

Planning of subsurface use

Birger Jansson and Torbjörn Winqvist

Swedish Building Research Summaries

S10:1976

Key words:
subsurface construction, rock cavity, tunnels, subsurface, urban planning, responsibility of planning, building control, inventory of establishments

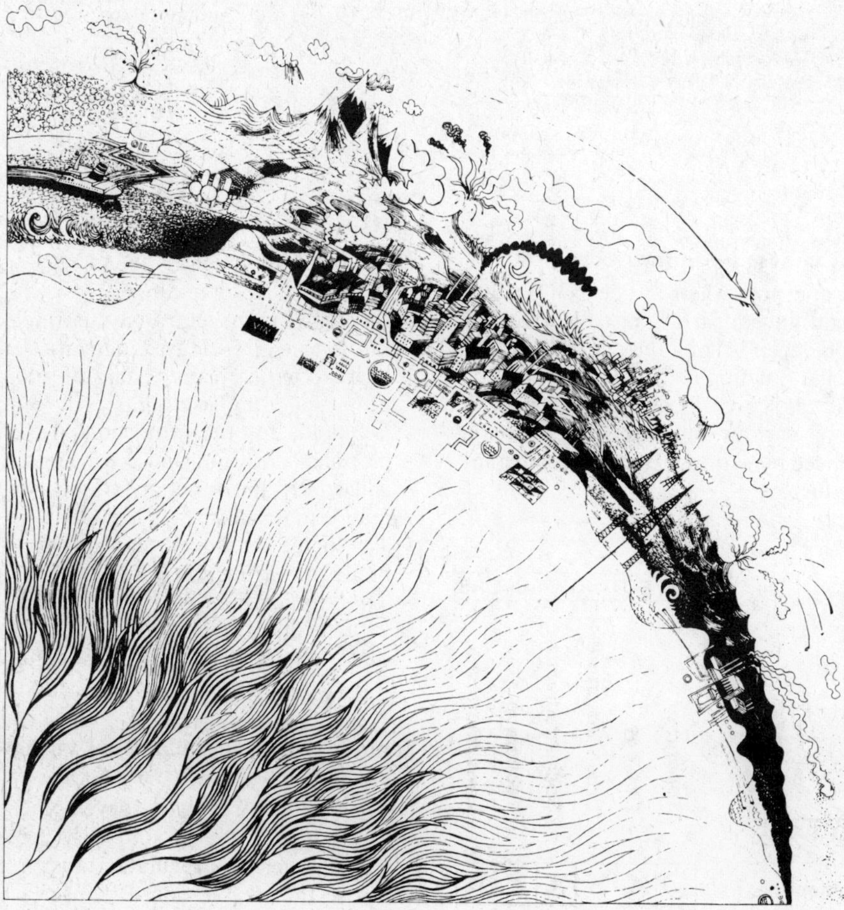

Tunnel and rock cavity construction has gone through a rapid development in Sweden. A highly developed tunneling technique has made it possible to lower costs. However, society's knowledge of the significance and consequence of tunneling is extremely limited, and does not at all match the advanced technology which is a distinctive feature of present subsurface construction. At present, there exists no anticipatory planning as regards subsurface areas.

Tunneling – an important resource in urban construction work – would be more efficiently pursued

- *if existing establishments were registered*
- *if principles as to the co-ordination of the needs of different parties were developed*
- *if the matter of responsibility for planning and control of subsurface establishments were made clear.*

An investigation team with representatives from various professional areas, assisted by a reference group, has tried to analyze the subject and develop overall aspects as to how the problems should be tackled.

In order to form a basis of analysis subsurface construction work has been studied in eight municipalities of various size and geographic location. Detailed studies have been pursued in one urban and two suburban areas. An account of preliminary conclusions has been distributed to authorities and organizations concerned. The present report is a revised version of different points of view which we have received.

Summary S10:1976 refers to research grant 730591-8 from the Swedish Council for Building Research to VBB Vattenbyggnadsbyrån, Stockholm.

UDC 69.035.4:711
624.19
69.009
SfB A
(20)

Summary of:

Jansson, B & Winqvist, T, 1976, *Planning of subsurface use.* (The English version is to be published in the D-series from the Swedish Council for Building Research under 1976) and *Undermarksplanering* (the Swedish version is to be published in the T-series from the Swedish Council for Building Research under 1976). (Statens råd för byggnadsforskning.) Stockholm.

Summaries in Swedish and English are published separately by the Swedish Council for Building Research.

Distribution:
Svensk Byggtjänst
Box 1403
S-111 84 Stockholm
Sweden

Scope of subsurface construction

About 10 % of all construction costs (excluding buildings) in Sweden relate to tunnels and rock cavities (1974). About 5 % concern power generation, mainly hydro-electricity, and 3.5 % storage rooms, mainly intended for oil. The remaining 1.5 % concern service and communication establishments within or in close connection with urban areas. The annual addition of subsurface space (excluding mines) amounts to 4 Mm^3. As a comparison, it should be noted that the annual growth of subsurface space in mines is about 23 Mm^3

Various investigations show that subsurface construction (excluding mines) increases at a pace which will bring about a doubling in ten years' time. Internationally, the scope of subsurface construction in Sweden is remarkably large. In proportion to their population, Norway and Sweden are numbers 1 and 2 in the world.

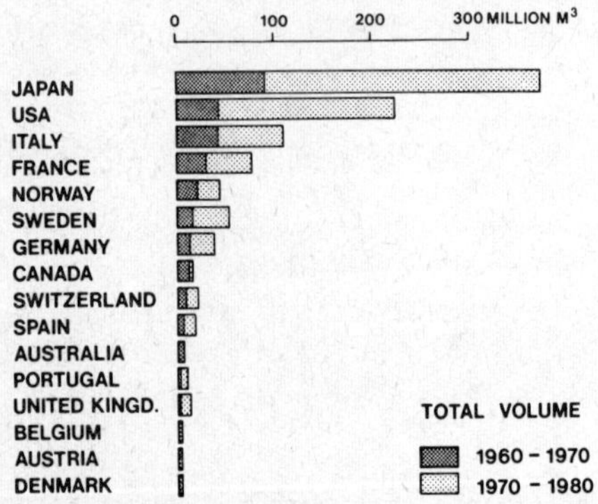

FIG. 1 Volume of tunneling in some OECD countries

Urban tunneling has been concentrated to certain parts of the country, where topography, geology and structure of built-up areas give cause for choosing subsurface positions. Stockholm and Gothenburg, the largest cities in Sweden, have the most intensive subsurface utilization in the country, but there are also small towns with a population of 15.000–50.000 which utilize considerable volumes of subsurface space. In many such towns, subsurface volume amounts to 10 m^3 per person.

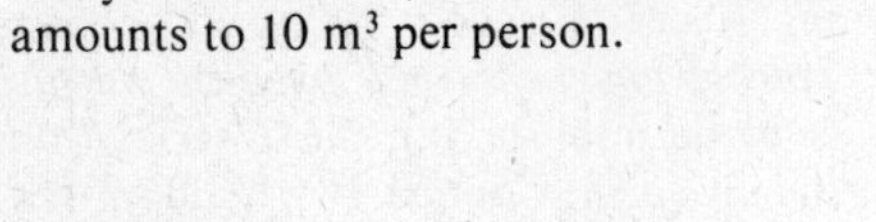

FIG. 2 Development of tunneling in Uddevalla town

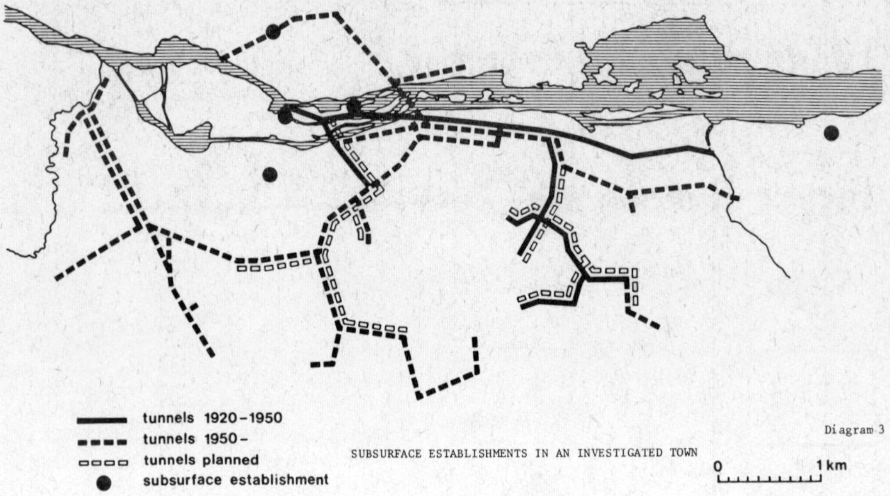

FIG. 3 Subsurface establishments in an investigated town

Why choose subsurface location?

The question, whether a certain function should have a surface or subsurface position depends upon the demands of that specific function.

The choice of location is often made up of various motives, which are influenced by various external circumstances—factors.

AFFECTING FACTORS

MOTIVES	Topography	Geology	Urban structure	Social demands	Technique	Laws and adm.
Connection	O	X	X	X	X	X
Protection	X	O	X	X	X	X
Space	X	X	O	X	X	X
Environment	X	X	X	O	X	X
Cost	X	X	X	X	O	X
Time	X	X	X	X	X	O

FIG. 4 Connection between motives and affecting factors

By quantifying and evaluating motives and affecting factors, a basis of cost – benefit analysis is formed, which may help in comparing alternative projects. In the same way, the "demand" for subsurface construction can be estimated as a part of planning.

Certain simple connections between motives and affecting factors are particularly conspicuous:

- A complicated topography will result in testing subsurface communications.
- A subsurface establishment in a high quality geological material has main advantages as regards the protection of the function in question.
- In a highly concentrated metropolitan area, an increased demand for space often enforces testing of additional space under ground.
- Demands for the protection of people and nature from disturbances, caused by the function in question, may lead to subsurface location.
- The highly developed tunneling technique may bring about lower costs.
- The juridical handling of subsurface establishments is simple, due to a limited right of disposal, and this may bring about a significant gain of time.

Tunneling demands a rational method of planning

Today, society regulates development and activities above ground very carefully. However, this regulation does not comprise the exploitation of subsurface areas to any great extent. Urban subsurface construction is very much dependent upon the "surface development" of society, though. Therefore, the planning of subsurface construction should be integrated into the overall planning, for which society, mainly the local authorities, is responsible.

The strong tie to functional needs and demands necessarily gives the initiative as to the testing of subsurface locality to those authorities which are responsible for the planning and realization of each particular function. e.g. state and local authorities, defence organizations, industries, power suppliers, storing companies and transport companies.

In order to utilize subsurface areas rationally, it is necessary to analyze thoroughly such aspects and fields of knowledge which are significant to subsurface construction. Therefore, we have seen as one of our main tasks to elucidate the existing amount of knowledge within the fields in question. In doing so, we were also able to provide a basis of pursuing a meaningful and methodical planning.

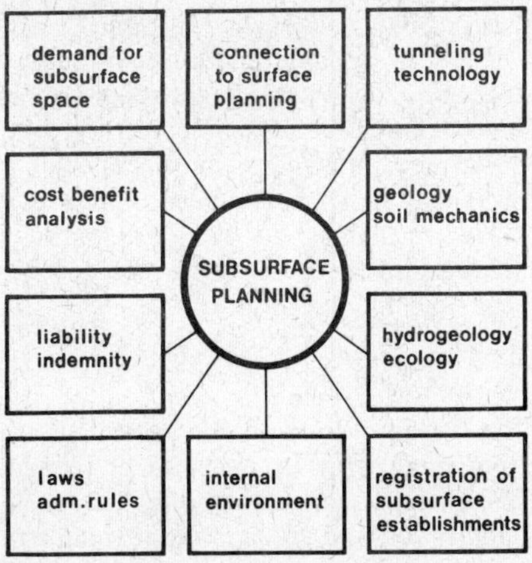

FIG. 5 Fields of knowledge connected with subsurface planning

The demand for subsurface space might be expressed by giving an account of preparatory projects which have reached high probability of completion. It is extremely hard to propose a reliable prognostication as regards long term development, since estimates of the future must be based partly upon functional needs, partly upon the inclination towards choosing subsurface location. This inclination is, as we have pointed out above, a compound of many local and general estimations. At present, there are no general statistics concerning subsurface construction. However, there is one research project. "Demand and Forecast", which is supported by the Swedish Council for Building Research (BFR), which aims at providing basic data for a continuous account of general statistics in these matters.

A connection between subsurface planning and surface planning can be achieved by incorporating subsurface planning into existing surface planning routines, which are of general significance. Society is responsible for overall planning concerning co-ordination and restriction of resources, e. g. ground, economy, labour, natural resources etc. At the same time, authorities responsible for particular functions pursue a specialized and expensive planning in order to meet the demands of society, e. g. transport, storing, water supply, refuse destruction, electricity, heating, telephone, defence establishments etc.

Overall planning provides starting points for the planning of particular functions. On the other hand, planning of particular functions provides a basis for overall planning. Planning means different degrees of foresight:

- long term planning in order to account for purpose, programme and reservations
- medium term planning in order to establish the need of expansion schemes
- short term planning aiming at completion and building

In close connection with the time aspect of planning, there are aspects of physical extension and degree of detail, in which three branches may be distinguished:

- outline planning, covering a whole municipality or region
- partial planning, covering parts of a municipality or system
- detail planning, which comprises objects or blocks and which may be co-ordinated with the designs

Demands for co-operation in planning enforce long term estimates of subsurface utilization, especially among bodies responsible for particular functions.

Tunneling technology. This report shows that it is possible to work out estimates of cost of construction work, to be used in planning. We have collected experiences concerning excavation, different types of reinforcements and various additional costs (transports etc.). Cost relations between conventional drilling and blasting and different special methods are accounted for, as is construction in other media than primary rock.

The account of technical conditions should also include bases of describing what consequences the building activities may have in surrounding areas.

Geology and soil mechanics are basic fields of knowledge as regards subsurface planning. Planning on various levels designing and building must be based upon adequate investigations. Advanced methods and first class equipment have been developed, which will help to carry out necessary investigations at various phases. However, investigation costs are increasing severely, concurrently with the demands for detailed information. Therefore, these costs must be weighed against the need of information and against estimated risks of making erroneous conclusions. In the report, principally necessary information for various phases in planning is pointed out.

The research project "Value and Scope of Preinvestigation" to which the Swedish Council for Building Research (BFR) contributes economically, has a very great value as regards relating methods of investigation to observations, made during the time of building, and indicating a method of investigation which will provide the desired information in the planning phases.

Hydrogeology. Subsoil water is often influenced by subsurface building and it is, furthermore, an element of risk, particularly during the building phase. Changes in groundwater balance is the most common cause of disturbances and damages in areas surrounding subsurface establishments, damages, which may, furthermore, be extremely extensive. Thorough studies of this kind of problems during recent years have led up to methods which make it possible to anticipate and describe changes and to take countermeasures, if necessary. Analyses of groundwater conditions are very important factors in the overall consequence analysis of a subsurface establishment.

Registration of subsurface establishments and geological data are of great importance as regards providing information for subsurface planning and subsurface construction. Complete establishments provide the most valuable information possible about geo-conditions and this information must therefore be taken care of for future needs. The position of subsurface establishments and information about the foundation of

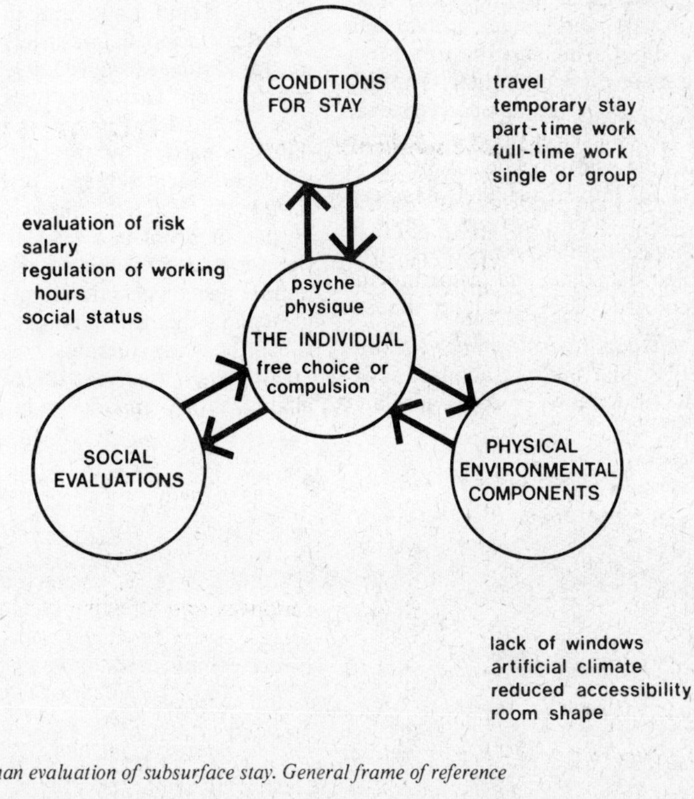

FIG. 6 Human evaluation of subsurface stay. General frame of reference

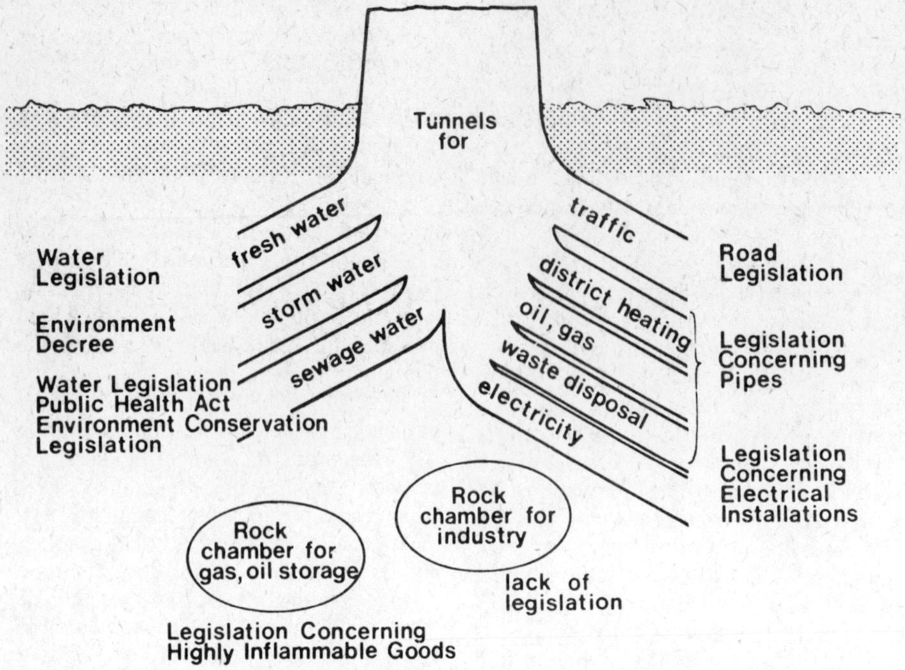

FIG. 7 Current legislation for subsurface use

surface buildings should be mapped in a special atlas for the benefit of both surface and subsurface planning.

Current practice as regards secrecy in military matters etc. should be critically examined in order to avoid withholding important information from the planners. (This is quite possible under current law.)

Internal environment. Environmental aspects of working and staying in subsurface establishments are of great importance. In the report, an inventory of what is known about human reactions during stay in subsurface establishment is accounted for. Up to the present day, this field of knowledge has not been satisfactorily dealt with. No physiological reactions have been found. However, certain social and psychological reactions towards work under ground can be pointed out, which are due to the fact that there are no windows etc. The risk of catastrophe should be observed, evaluated and reduced in order to counteract psychological blocking. A frame of reference for human reactions has been set up, which should serve as a guide in more thorough plannings, where environmental demands and demands for work democracy are given more space.

Laws and administrative rules According to the valid interpretation of existing laws, a landowner owns his land from the ground surface to the centre of the earth. In practice, however, his right of disposal is limited. Through voluntary agreements or agreements settled in court, an applicant may get the right to carry through a subsurface project without compensating the landowner. Furthermore, an individual person's right to carry on underground building may be juridically limited by officially sanctioned regulations.

A number of laws regulate the functional use of subsurface space. On the other hand, there is no legislation regulating planning, construction and control of subsurface cavities. Furthermore, there is no explicit obligation for local building committées or any other authorities to answer for this.

Liability and indemnity. The Swedish Code of Land Laws and the environment legislation dictate strict responsibility for damages caused by subsurface construction. However, there exists no precedential legal usage as to yielding damages caused by lowering of the water table. Such damages often affect a third party. Since the third party has the burden of proof in a trial, it is in practice very hard for him to get compensation. House owners do not get insurance protection against damages caused by subsurface constructions.

Cost-benefit analyses are of great significance for comparisons both between surface and subsurface establishments and between different kinds of subsurface establishments. In the report, different examples of analytic method have been accounted for. Consequence analysis may either be included in the cost–benefit analysis, or added separately, depending upon the degree of detail of the object in question.

Recommendations

The result of the research project shows that there is an urgent need of reform, particularly concerning official responsibility for, and administrative treatment of subsurface construction. This need can be summarized in the following recommendations:

Legislation or directions from the National Board of Urban Planning should comprise:

1. Demands for the registration of subsurface establishments
2. Demands for the registration of geoinformation from preinvestigations and from complete establishments
3. Demands for the planning of subsurface construction, which should comprise outline planning, partial planning and detail planning
4. Demand for obligatory building permit for subsurface building
5. Demands for obligatory geo-analysis, description of consequences and cost–benefit analysis in all preparatory examination of subsurface construction
6. Demands for continuous checking-up and control of subsurface establishments
7. Department responsibility for subsurface construction
8. Official expert institute, to which issues of subsurface construction are referred for consideration

Other measures should be taken to inform the parties concerned about the technique and planning of subsurface construction. Furthermore, education in adequate subjects should be provided to planners and technicians.

For the benefit of planning, problems concerning secrecy in military matters should be thoroughly studied, through the agency of the authorities.

A clear-cut legislation should strengthen the juridical protection of parties subjected to damage. Also, the possibility of introducing insurances, valid in these matters should be investigated.

Utgivare: Statens råd för byggnadsforskning